REAL-TIME STATISTICAL PROCESS CONTROL

Paul C. Badavas

The Foxboro Company
Foxboro, Massachusetts

PTR Prentice Hall
Englewood Cliffs, New Jersey 07632

Library of Congress Cataloging-in-Publication Data

Badavas, Paul C.
   Real-time statistical process control / Paul C. Badavas.
      p.   cm.
   Includes bibliographical references and index.
   ISBN 0-13-763574-5
   1. Process control—Statistical methods. I. Title.
   TS156.8.B34   1993
   658.5′62′015195—dc20   92-765
                                CIP

Editorial/production supervision
  and interior design: *Harriet Tellem*
Cover design: *Barbara Clay*
Prepress buyer: *Mary McCartney*
Manufacturing buyer: *Susan Brunke*
Acquisitions editor: *Michael Hays*
Editorial assistant: *Dana Mercure*

The publisher offers discounts on this book when ordered
in bulk quantities. For more information, write:

   Special Sales/Professional Marketing
   Prentice-Hall, Inc.
   Professional   Technical Reference Division
   Englewood Cliffs, New Jersey 07632

Printed in the United States of America

10 9 8 7 6 5 4 3 2 1

ISBN 0-13-763574-5

Prentice-Hall International (UK) Limited, *London*
Prentice-Hall of Australia Pty. Limited, *Sydney*
Prentice-Hall Canada Inc., *Toronto*
Prentice-Hall Hispanoamericana, S.A., *Mexico*
Prentice-Hall of India Private Limited, *New Delhi*
Prentice-Hall of Japan, Inc., *Tokyo*
Simon & Schuster Asia Pte. Ltd., *Singapore*
Editora Prentice-Hall do Brasil, Ltda., *Rio de Janeiro*

To three people I love very much,
my wife, Judith, and our children, Christos and Elizabeth.

Contents

Preface

The central purpose of this book is to demonstrate how to use real-time statistical process control (SPC) in the process industries in order to maintain and improve quality and increase productivity.

The scope of the book is to present the appropriate SPC tools and show how to use them for the actual operation of process plants. The main features of the book cover the following:

1. The fit of statistical process control with other operational process functions, including traditional process control.
2. How real-time SPC charts use the electronically collected and distributed real-time database for real-time monitoring and for on-demand analysis and operation.
3. How to take into account process dynamics in forming subgroups, and how to determine the autocorrelation of samples and the subgrouping method to use to minimize it. Plotting charts with minimum autocorrelation is necessary, because then the rules indicating the statistical control state of a variable can be applied with more confidence.
4. How to organize a process/plant using the cause and effect hierarchy and how to use the hierarchy for actual process/plant navigation and real-time monitoring.

5. How to analyze the process/plant and how to set up real-time SPC for actual operation.

6. The additional functions required by real-time SPC for on-demand analysis and real-time monitoring of process/plant operation.

7. The application of the unsymmetrical version of the CUSUM Chart, developed by the author, for variable drift detection.

8. The application of the CUSUM Controller and the Optimum Setpoint Controller for automatic closed-loop control. These controllers were invented by the author.

9. How to approach the implementation of real-time SPC in the oil and gas industry, the chemical industry, the pulp and paper industry, the food industry, and the minerals industry.

10. Overview of the statistical quality control/statistical process control management philosophy and its use with real-time SPC charts and tools to solve process/plant problems.

11. When, why, and which charts and rules to use for on-demand analysis and real-time monitoring of cause and effect variables.

12. When, why, and which charts and rules to use for on-demand analysis and real-time monitoring of attribute variables.

13. How and why to use combinations of official (fixed) and calculated values of the mean and standard deviation to evaluate the rules.

14. For each SPC chart, all calculations and a numerical example with the resulting chart and calculations.

The general plan of the book is first to discuss the relationship of real-time SPC and other operational process plant functions, including traditional process control. Second, the book presents the underlying theory, calculations, and statistics for individual and subgrouped samples. Some of the topics discussed are normality and measures for it, reasons for subgrouping data, the meaning and rules of statistical control, and mathematical and other useful transformations.

Then the discussion focuses on process dynamics and how to determine the autocorrelation for contiguous samples of a dynamic variable. In particular the author shows how to subgroup these samples in order to minimize their autocorrelation. The resulting charts can then be applied with increased accuracy in determining the statistical control state of a process variable.

Next, the book provides a complete description of the SPC charts and other tools, including innovations and new inventions made by the author. For each SPC chart, the book provides all calculations and a numerical example, with the resulting chart and calculations and its use, and other pertinent material.

Using the preceding sections as the foundation, the discussion then centers on the important issues of how to analyze, organize, set up and operate a process/ plant with real-time SPC. Chapters 15 through 18 provide this information.

Chapter 15 describes how to organize a process/plant with the cause and effect diagram and the cause and effect hierarchy, and how to use the hierarchy for actual process/plant navigation and real-time monitoring. Chapter 16 presents additional functions required by real-time SPC for on-demand analysis and process/plant operation.

Chapter 17 shows how to analyze the real-time database and how to choose the appropriate charts and other tools to set up real-time SPC for operation. Chapter 18 then shows how process/plant operators can use real-time SPC for on-demand analysis and real-time monitoring.

Next, the book presents an overview of the statistical quality control/statistical process control (SQC/SPC) management philosophy and its use with real-time SPC charts and tools to solve process/plant problems.

Chapters 20 through 24 describe how to approach the implementation of real-time SPC in the oil and gas industry, the chemical industry, the pulp and paper industry, the food industry, and the minerals industry. Everything is now in place to take advantage of real-time SPC to operate plants in the process industries to continuously maintain and improve quality and increase productivity.

ACKNOWLEDGMENTS

I would like to express my thanks and appreciation to David W. Noon for introducing me to the subject of statistical process control. Also, I acknowledge my debt to Dr. Albert D. Epperly, who coauthored with me some of the first papers on SPC referred to in this book and who has always been available for discussion and constructive criticism. My thanks and appreciation to Victor R. Tulli, who has been a constant source of help and encouragement. I would also like to express my gratitude to Elizabeth Naylor-McDevitt, Manager Applications Software Development, for her continuous support.

Chapter 1

Control Systems
and Real-time SPC

Statistical quality control (SQC) is an approach used to maintain and improve quality that results in improvement of productivity. The primary tool of SQC is *statistical process control* (SPC), which emphasizes the constant use of data and their interpretation during processing to determine the causes of variation in the final product and to make adjustments to improve it. Thus SPC is used to monitor the production process and make timely modifications to process variables to improve and/or maintain quality and increase productivity.

SQC/SPC is both a management philosophy and a set of tools. Reference 1 provides a good discussion. An overview of the management philosophy is given in Chapter 19 to provide management perspective for the use of the SPC tools.

This chapter first discusses the relationship of SPC and other operational process functions, including traditional process control. A control example is then used to introduce control and its terminology and to define the types of variables used throughout the book. Then the discussion covers the integration of SPC with control systems and provides a definition for real-time data and their collection into what is referred to as the *real-time database*. Finally the subject turns to monitoring with appropriately built charts and problem solving.

1.1 SPC AND OTHER OPERATIONAL PROCESS FUNCTIONS

Figure 1-1 shows where SPC fits in relation to other process plant operations or facility management functions, such as planning, performance, and control. Program logic and discrete control are used to automatically control the mechanical industries, for example, automobile manufacturing. Batch management and control are used for automatic recipe management and control of batch processes, such as batch chemical reactors and many batch processes in food making. Basic and advanced regulation is used to automatically control continuous processes, for example, oil refining. These three types of control will be referred to as *traditional control*.

SPC is used for the analysis and monitoring of any or all the process variables for all these processes. In general, SPC helps process operators and operations management to decide what process adjustments to make. Thus the operator closes the control loop; that is, control is not automatic. This type of control is called *open-loop advisory*.

The functions above SPC, such as raw material and production control, also

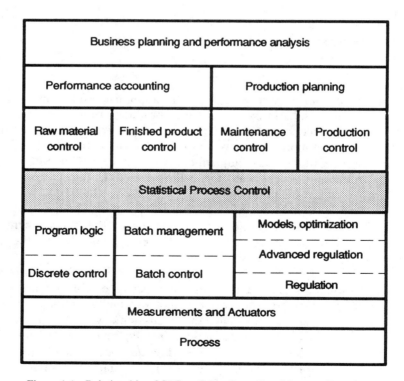

Figure 1-1 Relationship of SPC to Other Operational Process Functions.

benefit by the use of SPC tools, because such use improves understanding and can result in the identification of potential improvements.

1.2 CONTROL EXAMPLE AND TYPES OF VARIABLES

At this point, a simple control example is presented in order to introduce control and its terminology. The example is also used to define the types of variables used throughout the book.

Consider a shell and tube heat exchanger whose purpose is to heat cold water to a desired temperature by condensing steam. The cold water flows continuously through the tubes of the exchanger, while steam enters the shell side. Heat is transferred from the steam to the cold water through the tubes, and the condensed steam, now water called condensate, is continuously discharged from the bottom end of the heat exchanger shell.

The simplest way to automatically control the hot water temperature is to measure it and use it with a temperature controller that automatically throttles the steam valve to maintain the hot water temperature at a desired value. The hot water temperature is referred to as the *controlled variable*, the valve position is referred to as the *manipulated variable*, and the desired hot water temperature is referred to as the *set point*.

The amount of steam needed depends on steam pressure, cold water flow, and cold water temperature. These variables are referred to as the *load* or *disturbance variables*. A change in any load or disturbance variable results some time later in a change in the hot water temperature away from its set point. As the hot water temperature changes, the controller uses the change to throttle the valve in a direction that forces the temperature toward the desired set point. This control action takes place after the temperature changes, that is, after the fact, and is referred to as *feedback control*.

The steam pressure variation effect can be eliminated by adding a steam flow controller that uses a steam flow measurement and it, instead of the temperature controller, manipulates the steam valve to maintain steam flow at a desired set point. The temperature controller then manipulates the steam flow set point to control hot water temperature at the desired set point. This control is termed *cascade control* in that the steam flow controller is cascaded to the hot water temperature controller.

The control can be improved further by minimizing the effect of cold water flow and temperature changes. This is done by calculating a heat balance for the exchanger that provides a relationship that computes the flow of steam needed as a function of cold water flow and temperature. This is termed *feedforward control*, because it changes the manipulated variable as soon as the load variables change, as opposed to waiting for the hot water temperature to change and then taking feedback action, as previously. The feedback controller is still needed, however, because the heat balance is not perfect. In this case, the feedback control action is very small, and is referred to as *feedback trim*, in comparison to its action without the heat balance feedforward control.

In subsequent discussion, a variable is under *control* or *controlled* if it is automatically controlled by a controller such as the hot water temperature and steam flow controller described above. This control action is referred to as *closed-loop control*. The steam flow measurement, the steam flow controller, and the steam valve are referred to as a *control loop*, or *loop* for simplification. The combination of the hot temperature controller and steam flow controller constitutes a *cascade control loop*, whereas the combination of the heat balance and cascade control loop make up the *feedforward–feedback control loop*.

Controlled, manipulated, and set point are the three types of variables discussed so far. *Effect variables* are associated with the product(s) or by-product(s), such as the hot water temperature in the above example. And variables associated with feeds to a unit, such as the cold water flow and temperature and steam flow and pressure, as well as variables associated with the unit itself, such as the heat exchanger tube volume and shell pressure, are referred to as *cause* or *causal* variables.

The relationships between cause and effect variables are referred to as *causal relationships*. An example is the causal relationship between hot water temperature and steam flow, which states that the hot water temperature increases with an increase in steam flow provided that the hot water temperature is not under closed-loop control and the cold water flow and temperature remain constant.

Another distinction made is the reference to *quality variables* for such variables as feed and product composition, moisture, viscosity, and density that are indicators of quality characteristics, as opposed to variables such as pressure temperature and residence time.

Another type of variable is the *attribute variable*, or *attribute* for simplification. Examples are the number of defective items and fraction defective and the number of defects and defects per unit. Charts associated with the attribute type of variable will be referred to as *attribute charts*, whereas charts for all other types of variables will be referred to as *charts for variables*.

Figure 1-2 shows typical responses of the hot water temperature to a change in cold water flow using the feedback cascade control and feedforward–feedback control. The cascade control is less effective, because the controller does not start to respond to the change in cold water flow until the hot water temperature changes. This takes time, because it takes time to displace the volume of water existing in the tubes. This causes the temperature response to follow a damped cycle, that is, one that gets smaller as time goes on. This is referred to as the *dynamic* or *transient response*.

The feedforward–feedback control combination improves the transient response, because the steam flow is changed as soon as the cold water flow changes based on the heat balance. Of course, there is some transient response, because of the dynamics that cannot be perfectly compensated for. This control is more efficient, because it requires less energy during the transient part of the response.

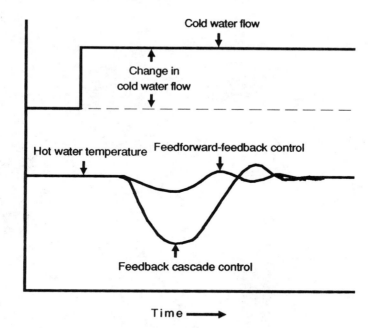

Figure 1-2 Typical Temperature Control Responses.

It does not come for nothing, though, because it requires the installation of additional measurements and computation.

If there are no other disturbances, the temperature settles near the set point, and this is referred to as its *steady state value*.

In practice, processes and their variables never reach steady state, because, in addition to changes in systematic load variables, there is also variation due to random disturbances, such as weather changes that affect the heat exchanger when it is not housed in a controlled atmosphere. Moreover, multiple load upsets can occur at different times. Therefore, all process variables exhibit variation due to changes in systematic load variables and also due to random disturbances.

The main purpose of a process control system is to provide safe operation, to provide stable control, to maintain selected process variables as close as possible to desired values, to save energy, and, when necessary, to operate against process constraints to increase throughput.

1.3 SPC WITH CONTROL SYSTEMS AND THE REAL-TIME DATABASE

State of the art control systems for the process industries and manufacturing plants are distributed functionally and geographically and are connected by communications networks, as presented in references 2 and 3.

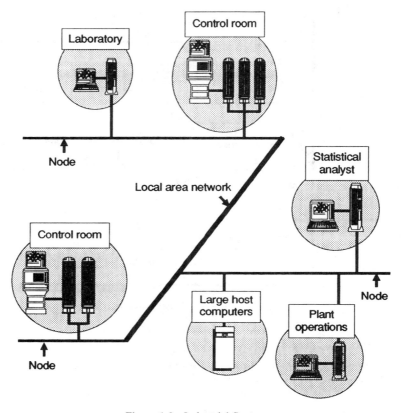

Figure 1-3 Industrial System.

Figure 1-3 shows an industrial control system that connects a number of plant areas: process input/output devices and control, workstation and application processors with the required bulk storage, input and output devices, and host computers connected to nodes. The nodes are connected by a local area network. The local area networks can also be connected to provide all necessary communications for a plant site.

The functions involved in the integration of SPC with traditional process control are shown in Fig. 1-4. These functional modules consist of the process input/output, the control, which contains continuous, sequence, and logic functions, the real-time database, SPC calculations, and the human interface.

The process input devices convert continuous analog signals from the process, such as flow, temperature, and pressure, into digital signals for computer processing. The variable signals are sampled on a periodic basis. The time between samples, the *sample period*, is fast enough so that timely control action can be taken to maintain controlled variables at desired set points, as well as for other

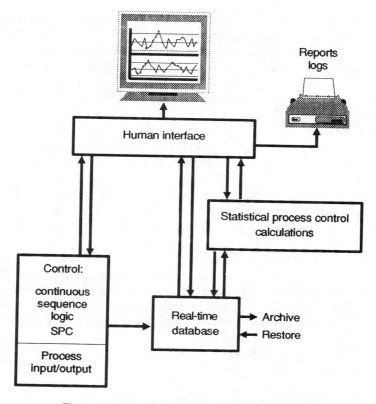

Figure 1-4 Functional Distribution within System.

actions. The input devices also sample discrete variables, such as on/off states of contacts and other logical input devices.

The samples of the input variables are then used by the continuous, sequence, and logic control functions, which eventually compute the required values for the output variables. The output devices convert the digital signals from the control functions to analog and discrete contact and other logical output signals that drive the actual manipulated process variables, such as valves, pumps, and motors, to achieve the desired control.

The real-time database automatically collects data from the control and input/output functions and stores them in on-line bulk storage, for example, disk drives. The same database also stores manually entered data by operators, laboratory analysts, and other plant personnel.

On-line storage for the real-time database should be months or years to provide for ready process analysis, operator training, reporting, and other functions. On-line storage combined with archive storage on electronic media should contain the real-time database for the lifetime of the process or manufacturing plant. Of course, all or parts of the archived real-time database should be easily restored for on-demand analysis and be accessible by the same tools that access the on-line real-time database. The real-time database is then used by the other traditional control functions shown in Fig. 1-1 and by the SPC charts and other tools.

SPC uses the following four types of information:

1. *Effect variables,* such as viscosity, composition, density, melt index, and brightness, used in xbar and range, xbar and sigma, individuals and CUSUM charts to monitor product quality.

2. *Cause variables,* such as flows, temperatures, pressures, and feed compositions, used in xbar and range, xbar and sigma, individuals, and CUSUM charts to monitor and determine the cause of poor product quality.

3. *Attributes,* such as fraction and number of defective items and number of defects and defects per unit, used in *P, NP, C,* and *U* charts to monitor end (final) product and overall process performance.

4. *Causal relationships,* which consist of text information. They provide concise and useful explanations of particular process characteristics that can be used on line to improve process understanding and operation. These also include operator-entered notes associated with a particular chart and time.

The two main sources of information are as follows:

1. *On-line sensors or other measurement devices*
   - For continuous variables, such as flow, level, pressure, and temperature, sampled at desired sample periods.
   - For sample data in numerical form, such as counters, weight cells, chromatographs, and automated smart sensors.
   - For discrete status, such as contact open/closed on switch on/off.

2. *Manually entered*
   - Manually entered numbers.
   - Manually entered notes and text.

In a real-time system, the data have the following attributes:

- Name
- Units, for example, gallons per minute or percent
- Date/time
- Validity status, for example, a bad data flag

These attributes are automatically adhered to in real-time systems. Reference to data by common name with engineering units at the point of interpretation is as important to SPC as it is to traditional process control. The date/time attribute

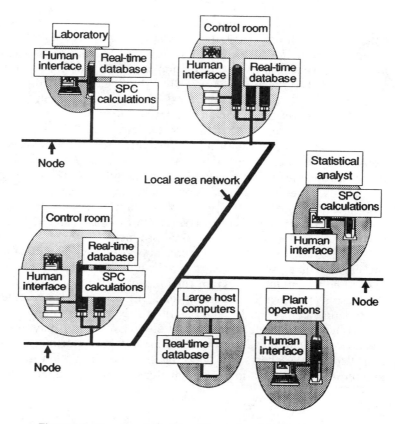

Figure 1-5   Functional Modules Allocated to Specific Processors.

of data is required for time correlation between data groups and for the determination of cause and effect.

The functional modules are distributed to run in different processors throughout the system. The process input/output and control run in control processors, while the real-time database modules run in application processors. The SPC calculation modules also run in application processors. The human interface modules run in workstation processors.

In an industrial system, a number of processors can be running simultaneously. Figure 1-5 demonstrates the functional distribution among a number of processors. It shows four real-time database modules, three SPC calculations modules, and five human interface modules. All three SPC calculations modules can access all four real-time databases. The same SPC calculations modules support all five human interface modules. This distribution of functions and modules among different processors connected by the communications network constitutes what is usually referred to as the *distributed control and information system.*

The term *real-time database* used throughout the rest of the book refers collectively to all the distributed databases that collect and store the value, date/time, and status for all plant variables from the current (present) time to some time in the past. It is important to emphasize the fact that the real-time database is always tied to the current time. Thus the SPC charts and other analysis tools, when requested, automatically access data from the current time backward. This way timely action can be taken to improve operation.

In general, real time means that microprocessors and computers used in distributed control and information systems can collect data, perform computations, and take action in a timely fashion, which is a function of both process dynamics and the action to be taken. For example, a flow controller has to respond in seconds, whereas a composition controller can take minutes or hours. Moreover, the final product may go through many intermediate steps that may take hours or days, whereas many intermediate variables can be acted on within hours and still affect the quality of the final product.

SPC charts and other SPC tools access the real-time database starting with the current time. When a chart is scheduled for periodic real-time monitoring, everytime the monitor program runs it retrieves enough data from the historical database starting with the current time. It then evaluates whether the variable is in statistical control and sets an alarm if it is not.

When the operator calls a chart on-demand for display, the display program retrieves enough data from the real-time database starting with the current time and displays the chart. The operator then has the option to move back in time and redisplay the chart in order to analyze the process operation and take action.

1.4 MONITOR WITH APPROPRIATELY BUILT CHARTS AND SOLVE PROBLEMS

Samples of a process variable are usually correlated to each other, that is, autocorrelated. Their autocorrelation is worst when they are sampled periodically at short sample periods in relation to the dynamics of the process where the variable comes from.

Subsequent chapters show how to take process dynamics into account and how to minimize the effect of autocorrelation in building SPC charts. The charts can then be used more effectively for process monitoring. This is because the rules used to monitor the in-statistical-control state of a process variable apply with greatly increased confidence to samples whose autocorrelation is made as small as possible. This increases the probability that when a rule is violated it is, indeed, due to assignable causes and the process needs attention.

The SPC charts can be used to monitor variables that are not controlled, as well as variables that are. Typically, a variable is not under closed-loop control because of lack of appropriate on-line sensors. Samples for the variable are obtained by laboratory analysis and entered in the real-time database whenever the analysis is completed.

A variable that is under closed-loop automatic control can also be monitored using the SPC charts. In this case, the rules and other parameters are chosen so that the rules are violated only when the variable strays more than an acceptable amount from its desired set point. This can also be done by charting the error, which is the difference between the set point and the measured value of the variable. This is referred to as monitoring the *range of controllability* of a variable and it is described in more detail in Chapter 5.

The SPC charts automatically check whether the monitored variables violate any of the selected rules that indicate when a variable is not in statistical control. When any rule is violated, the operator is automatically informed by the setting of an alarm or other means. The operator then selects charts for other variables that help him to track the cause of the problem and implement the necessary control action.

Typical control actions consist of the following:

- Changing appropriate variable set points or targets.
- Improving control of upstream units to minimize the introduction of systematic variation into the downstream process.
- Retuning the controllers and modifying associated control functions.

If operator actions are not sufficient to solve the problems, then operations management and engineering need to get involved to solve them.

Operations management and engineering also use the charts to decide when to make raw material, process equipment, and control system changes. For a given process, specific solutions to problems require deeper understanding of the physics, chemistry, engineering, mathematics, and other sciences.

The unit approach uses cause and effect diagrams for the individual units. It involves identifying and charting quality variables and measures of performance of the raw materials and products of the unit. To do this requires the identification of both internal and external customers and specifications for the products of each unit.

This approach is similar to the approach followed in the process industry for years. The differences, however, are as follows:

- Emphasizing quality and performance at all levels.
- Providing the necessary tools, knowledge, and understanding down to the level where the product is made and the process is adjusted.
- Using the appropriate SPC tools and traditional process control to deal with the process and its problems.
- Using real measures of performance to make decisions about the process.

- Emphasizing cause and effect all the way from raw materials to final products.
- Using the SPC tools to help to involve everybody in striving to improve product quality and performance continuously for the individual units and the whole plant or process.

Chapter 2

Statistics
and Individual Samples

This chapter presents the necessary statistical calculations for individual samples. In particular, it introduces the normal distribution and its equation, the individuals histogram, and calculated measures for the normality of samples. In addition, it presents mathematical transformations that can be used to achieve normality under certain conditions. Moreover, it provides the calculations for mean and sigma values for individual samples. Reference 4 covers some of this material in more detail. In addition, this chapter discusses the meaning of in statistical control and the rules used to check for it. Moreover, it provides some suggestions as to which and how many rules to use for real-time monitoring. Finally, it covers the possible combinations of calculated and official values of mean and sigma that can be used in evaluating the chosen rules.

2.1 MEAN AND SIGMA

Let $X(t1)$, $X(t2)$, . . . , $X(tNS)$ represent a group of individual samples from a real-time database, where $t1, t2, . . . , tNS$ are the time/date for each sample. For further simplification in notation, from now on let $X(t1) = X(1)$, $X(t2) = X(2)$, . . . , $X(tNS) = X(NS)$ denote the group of individual samples. Later the group of samples will be divided into subgroups of size N. To provide consistent notation throughout, let NS denote the number of subgroups. For individual samples, the subgroup size $N = 1$, and in this case the number of subgroups is equivalent to the number of samples in the group.

Two important measures computed from the data are the mean and standard deviation. The mean represents the average value or central tendency of the data, whereas the standard deviation is a measure of their dispersion. Typically, the standard deviation is denoted by the Greek letter sigma. The word *sigma* will be used from now on instead of the term standard deviation to simplify the notation of the computed variables that follow.

The mean value is computed as follows:

$$XB = \frac{\sum_{i=1}^{NS} X(i)}{NS} \tag{2-1}$$

The sigma can be computed two ways as follows:

$$S = \sqrt{\frac{\sum_{i=1}^{NS} [X(i) - XB]^2}{NS - 1}} \tag{2-2}$$

$$SIGP = \sqrt{\frac{\sum_{i=1}^{NS} [X(i) - XB]^2}{NS}} \tag{2-3}$$

where

NS = number of subgroups
$X(i)$ = ith sample
XB = mean
S = sample sigma, calculated using $NS - 1$ samples
$SIGP$ = sigma prime, calculated using NS samples

This way of calculating sigma is referred to as the rms (root-mean-squared) method. As NS gets very large, the difference between S and $SIGP$ becomes negligible. $SIGP$ is always the smaller value.

2.2 NORMAL DISTRIBUTION

One of the best known and widely applicable distributions is the normal distribution. The equation of the normal distribution is

$$p(X) = \frac{1.0}{SIGP \sqrt{2\mathrm{pi}}} \exp\left\{\frac{-1}{2}\left[\frac{X - XB}{SIGP}\right]^2\right\} \tag{2-4}$$

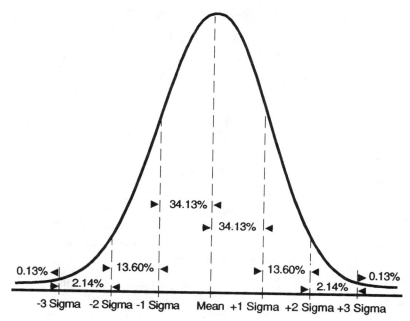

Figure 2-1   Normal Distribution.

where

pi = 3.14159

X = random variable whose range is from minus to plus infinity

Figure 2-1 shows the normal distribution, which is symmetrical about the mean value and its total area is 1.0 (100%). The area of the normal curve between −1 sigma and +1 sigma is 0.6827 (68.27%), between −2 sigma and +2 sigma is 0.9545 (95.45%), and between −3 sigma and +3 sigma is 0.9973 (99.73%).

2.3 INDIVIDUALS HISTOGRAM

A *histogram*, also called a frequency histogram, is a frequency distribution of a set of data. The total range of the sample values is divided into a number of cells. The count of samples (frequency) for a given variable whose value falls within the range of values represented by a cell is plotted against the value of the variable. The histogram can be plotted in many different ways. One is the bar graph. Each bar represents a cell. The bar length (height when plotted vertically) is proportional to the number of samples whose values fall within the range of values of the cell. The vertical bar graph is used to plot histograms in this book.

The following computations are also needed to build the histogram:

1. Select the number of cells, using the following table as a guide.

No. of Subgroups (NS)	No. of Cells
21–50	7
51–80	8
81–100	10
101–175	12
176–250	14
251–325	16
326–400	18
401–max	20

Histograms for less than 21 samples are not recommended, because the sample size is too small to draw any meaningful conclusions.

2. Find the maximum and minimum values in the group of samples and compute the class interval, CI, as follows:

$$CI = \frac{maximum - minimum}{number\ of\ cells} \tag{2-5}$$

Table 2-1 contains 100 samples for a variable named rang200. Their mean is $XB = 54.99$, and their sigma is $SIGP = 1.90$. Figure 2-2 shows the individuals

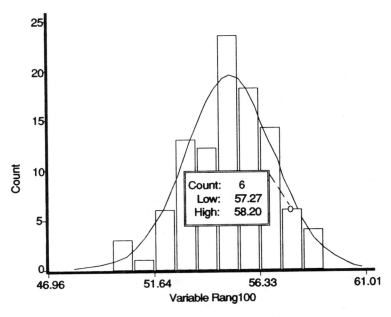

Figure 2-2 Histogram of Samples in Table 2.1.

TABLE 2-1 INDIVIDUAL SAMPLES

Sample	Date	Time	Value	Sample	Date	Time	Value
1	10/01	00:05:00	54.57	51	10/03	02:06:00	55.87
2	10/01	01:10:00	58.47	52	10/03	03:15:00	56.55
3	10/01	02:06:00	54.96	53	10/03	04:10:00	57.22
4	10/01	03:15:00	56.17	54	10/03	05:08:00	55.23
5	10/01	04:10:00	57.57	55	10/03	06:07:00	52.66
6	10/01	05:08:00	49.47	56	10/03	07:12:00	52.66
7	10/01	06:07:00	55.39	57	10/03	08:10:00	55.09
8	10/01	07:12:00	56.39	58	10/03	09:05:00	54.04
9	10/01	08:10:00	52.12	59	10/03	10:18:00	52.56
10	10/01	09:05:00	54.84	60	10/03	11:20:00	54.24
11	10/01	10:18:00	50.64	61	10/03	12:14:00	54.78
12	10/01	11:20:00	54.48	62	10/03	13:15:00	54.34
13	10/01	12:14:00	56.74	63	10/03	14:22:00	57.88
14	10/01	13:15:00	52.48	64	10/03	15:25:00	56.22
15	10/01	14:22:00	59.14	65	10/03	16:18:00	57.37
16	10/01	15:25:00	55.50	66	10/03	17:15:00	56.31
17	10/01	16:18:00	56.18	67	10/03	18:12:00	54.48
18	10/01	17:15:00	54.29	68	10/03	19:14:00	55.37
19	10/01	18:12:00	54.15	69	10/03	20:20:00	56.23
20	10/01	19:14:00	57.07	70	10/03	21:17:00	54.57
21	10/01	20:20:00	56.71	71	10/03	22:30:00	54.66
22	10/01	21:17:00	58.78	72	10/03	23:35:00	53.46
23	10/01	02:30:00	54.00	73	10/04	00:10:00	55.27
24	10/01	23:35:00	51.68	74	10/04	01:05:00	57.05
25	10/02	00:18:00	53.75	75	10/04	02:06:00	58.38
26	10/02	01:25:00	52.93	76	10/04	03:12:00	55.50
27	10/02	02:06:00	57.56	77	10/04	04:11:00	55.50
28	10/02	03:12:00	56.97	78	10/04	05:08:00	54.76
29	10/02	04:11:00	52.67	79	10/04	06:08:00	51.63
30	10/02	05:08:00	56.73	80	10/04	07:12:00	55.02
31	10/02	06:08:00	55.34	81	10/04	08:15:00	57.32
32	10/02	07:12:00	54.42	82	10/04	09:05:00	55.80
33	10/02	08:15:00	53.07	83	10/04	10:17:00	56.60
34	10/02	09:05:00	54.52	84	10/04	11:20:00	54.32
35	10/02	10:17:00	54.52	85	10/04	12:04:00	57.17
36	10/02	11:20:00	53.65	86	10/04	13:15:00	53.02
37	10/02	12:04:00	54.62	87	10/04	14:12:00	55.19
38	10/02	13:15:00	52.99	88	10/04	15:20:00	56.86
39	10/02	14:12:00	52.80	89	10/04	16:18:00	56.00
40	10/02	15:20:00	53.38	90	10/04	17:19:00	53.42
41	10/02	16:18:00	53.12	91	10/04	18:12:00	55.78
42	10/02	17:19:00	56.08	92	10/04	19:19:00	54.49
43	10/02	18:12:00	54.76	93	10/04	20:11:00	52.00
44	10/02	19:19:00	51.65	94	10/04	21:17:00	53.18
45	10/02	20:11:00	57.23	95	10/04	22:21:00	57.68
46	10/02	21:17:00	56.17	96	10/04	23:32:00	55.56
47	10/02	22:21:00	54.42	97	10/05	00:02:00	56.76
48	10/02	23:32:00	50.01	98	10/05	01:23:00	56.31
49	10/03	00:13:00	54.56	99	10/05	02:06:00	56.31
50	10/03	01:10:00	55.50	100	10/05	03:10:00	54.42

histogram for these samples and the superimposed normal curve using Eq. (2-4) and the preceding *XB* and SIGP values. Visual inspection of the histogram indicates that the samples form a reasonably normal distribution. Additional confidence about normality is achieved by obtaining objective measures for normality, which is the subject of the next section.

2.4 SKEWNESS AND KURTOSIS

The calculated values of skewness and kurtosis provide quantitative measures for normality.

Skewness

The third moment about the mean normalized by the third power of the sigma to make it dimensionless is a measure of skewness and is computed as follows:

$$A3 = \frac{\displaystyle\sum_{i=1}^{NS} [X(i) - XB]^3}{(NS - 1)SIGP^3} \tag{2-6}$$

A positive value for *A3* indicates positive skewness, while a negative value indicates negative skewness. $A3 = 0$ indicates no skewness.

Skewness is the degree of asymmetry of a frequency distribution usually in relation to the normal distribution:

Skewness > 0 indicates that the histogram has a longer tail to the right of its central maximum as compared to the left. In this case, the distribution for the samples is said to be skewed to the right or have positive skewness.
Skewness < 0 indicates that the histogram has a longer tail to the left of its central maximum as compared to the right. In this case, the distribution for the samples is said to be skewed to the left or have negative skewness.
The normal distribution is symmetrical about its mean and its skewness = 0.

Kurtosis

The fourth moment about the mean normalized by the fourth power of the sigma to make it dimensionless is a measure of kurtosis and is computed as follows:

$$gamma = \frac{\displaystyle\sum_{i=1}^{NS} [X(i) - XB]^4}{(NS - 1)SIGP^4} - 3 \tag{2-7}$$

For the normal distribution, gamma = 0. gamma < 0 indicates a leptokurtic distribution. gamma > 0 indicates a platykurtic distribution.

Kurtosis is the degree of peakedness of a frequency distribution, usually in relation to the normal distribution:

Kurtosis < 0 indicates a thin distribution with relatively high peak, and it is called leptokurtic.

Kurtosis > 0 indicates a distribution that is wide and flat topped, and it is called platykurtic.

The normal distribution that falls between these two distributions is called mesokurtic, with kurtosis = 0.

Assuming that a large enough number of samples is used in the calculations (at least greater than 20), a good rule of thumb is to accept the data as normally distributed when

$$-1.0 \leq \text{skewness} \leq 1.0 \tag{2-8}$$

$$-1.0 \leq \text{kurtosis} \leq 1.0 \tag{2-9}$$

The skewness and kurtosis of the samples used to construct the histogram of Fig. 2-2 are

$$\text{skewness} = -0.33 \tag{2-10}$$

$$\text{kurtosis} = -0.03 \tag{2-11}$$

These values for the normality measures furnish objective verification for the visual indication that the samples of Table 2-1 are normally distributed according to the rule of thumb stated above.

2.5 MATHEMATICAL AND OTHER USEFUL TRANSFORMATIONS

When the variable itself is not normally distributed but a function of the variable is, a mathematical transformation on the variable samples can be performed. For example, a log-normal distribution is transformed to normal via the natural log function. The logarithmic transformation is

$$Y = C1 \ln(X) \tag{2-12}$$

$C1$ is a parameter that can have any positive or negative value. Applying this transformation with $C1 = 1.0$ to the samples in Table 2-1 results in skewness = -0.43, which is more negative than the skewness of -0.33, the skewness without any transformation. Thus the logarithmic transformation skews data negatively.

TABLE 2-2 TRANSFORMATIONS, SKEWNESS, AND KURTOSIS FOR THE SAMPLES OF TABLE 2.1

Transformation	Skewness	Kurtosis
None	−0.03	−0.33
$Y = \ln(X)$	−0.43	0.0
$Y = \exp(0.1X)$	0.18	−0.20
$Y = \exp(-0.1X)$	0.88	0.96
$Y = X^{1/2}$	−0.38	0.02
$Y = X^2$	−0.23	−0.12
$Y = X^{-1/2}$	0.47	0.16
$Y = X^{-2}$	0.63	0.41

Other useful transformations are the exponential and the variable raised to a power. Their equations are

$$Y = C1 \exp(C2\ X) \qquad (2\text{-}13)$$

$$Y = C1(X)^{C2} \qquad (2\text{-}14)$$

$C1$ and $C2$ are parameters whose values can be any positive or negative values, including fractions.

Table 2-2 summarizes the results of applying the above transformations to the samples of Table 2-1. The results of Table 2-2 show that positive powers greater than 1.0 result in positive skewness, while positive powers less than 1.0 result in negative skewness. All negative powers result in positive skewness. The basis of comparison is the skewness of −0.33, which is obtained without any mathematical transformation.

The effect of using transformation $Y = \exp(-0.1\ X)$ is shown by the histogram of Fig. 2-3, which is positively skewed as compared to Fig. 2-2. Many other transformations can be used in addition to the ones given; however, these are a good set to start with.

Therefore, the histogram and the computed values of skewness and kurtosis are used to check which is the best transformation to use to approach normality for a given set of data. This is done by building histograms for each transformation type and selecting the transformation that forces the distribution to be closest to the normal distribution.

In many cases throughout industry, it is necessary to build charts that display the ratio of two variables. This is readily achieved by being able to configure charts that access two variables from the real-time database, form their ratio, perform the necessary calculations, and rule check and display them. For example, the ratio of chemical to raw material is a commonly controlled variable to maintain product quality. Another example is the ratio of catalyst to feed in a chemical reactor.

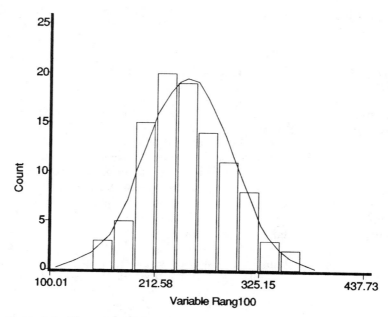

Figure 2-3 Histogram of Samples in Table 2.1 Using Transformation $Y = \exp(-0.1X)$.

In textile and paper manufacturing and other industries that produce rolls or sheets of materials, variables are measured across the sheet. For example, the pressure may be measured at four different points across the textile web. Now consider the need to chart the average of the four pressures. This is accomplished by accessing the four pressure variables from the real-time database, merging them into subgroups of four, performing the required calculations, and displaying them. Thus it is necessary to be able to configure a chart and merge a number of variables.

The transformation given by Eq. (2-14) can be used to convert a variable to other units. For example, pounds of steam per hour can be converted to dollars per hour using constant $C1$, where $C1$ is the value of dollars per pound of steam and $C2 = 1.0$. Displaying the chart with this transformation shows the operator the energy cost per hour for the process.

Another useful transformation is the multiplication of a variable by a constant and then adding a constant. This provides for span and bias adjustment for the displayed variable.

Based on their experience with specific processes and their variables, readers will see the need to use other types of transformations and data conversions to achieve particular objectives. The fact to remember is that the real-time database collects the raw data, but real-time SPC needs to provide the required transformations as part of configuration and display to facilitate process analysis and operation.

2.6 IN STATISTICAL CONTROL AND RULES FOR IT

A variable is *in statistical control*, when all assignable or special causes of variation have being eliminated and the remaining random variation is due to what is referred to in reference 1 as common causes. Based on the theory of runs, a number of rules have been developed and are used to check whether a variable is in statistical control. Table 2-3 lists 11 of these rules. Reference 5 is the main reference for the rules, while reference 4 contains some discussion. The rules are also discussed in several places throughout the literature on statistical quality control. The rules apply in general; however, the probability of a given rule being violated depends on the autocorrelation of the samples, which typically causes rules to be violated more frequently.

Reference 5 provides the following results for the probability of certain rules being violated when the distribution is nearly normal. The probability of rule 1 of Table 2-3 being violated is 0.27%, for rule 3 it is 0.30%, and for rule 8 it is 0.39%. Thus, when all three rules are used for real-time monitoring, the probability of any one being violated is their sum, which is 0.96%. This says that even though all assignable causes have been eliminated and the process is in statistical control there is still a probability of 0.96% that one of the three rules will be violated. This requires prudence in choosing which and how many rules to use and how to interpret them. The author recommends that no more than two of the rules of Table 2-3 be used for monitoring a given chart. In most cases, rules 1 and 3 are sufficient.

Rule 9, stratification, should not be used when the charted variable is under closed-loop control because, if the controller does its job and the load upsets are not continuous, which is usually the case, the end result is stratification. Rule 10, mixture, does not apply in usual control situations because mixtures come from more than one unit, and control is typically on a per unit basis. Rule 3 is similar to rule 4 and, therefore, one of them is sufficient.

Instead of using charts with rules 5 and 6, gradual upward/downward change in level, rules 7 and 8, upward/downward trend, and Rule 11, sudden jump in level, the author recommends the use of the CUSUM chart presented in Chapter 9.

2.7 CALCULATED AND OFFICIAL VALUES FOR RULE EVALUATION

The rules used to determine the in-statistical-control state of a variable listed in Table 2-3, as well as the control limits, require the central line mean and sigma values for their evaluation and calculation. The mean and/or sigma value used

TABLE 2-3 STATISTICAL PROCESS CONTROL RULES

Freak: A single point greatly different from the others.

> *Rule 1:* One point outside ± 3 sigma of the central line.

Erratic pattern: Large fluctuations that are wider than the control limits. Erratic up and down movement of points.

> *Rule 2:* Three consecutive points jumping ± 3 sigma or more.

Grouping or bunching: A number of points that are close together.

> *Rule 3:* Two of three consecutive points above $+2$ sigma or below -2 sigma from the central line.
>
> *Rule 4:* Four of five consecutive points above $+1$ sigma or below -1 sigma from the central line.

Gradual upward change in level: Gradually points shift upward to a new level.

> *Rule 5:* Eight consecutive points above the central line.

Gradual downward change in level: Gradually points shift downward to a new level.

> *Rule 6:* Eight consecutive points below the central line.

Upward trend: A number of points continuously increasing in value.

> *Rule 7:* Five consecutive points increasing in value.

Downward trend: A number of points continuously decreasing in value.

> *Rule 8:* Five consecutive points decreasing in value.

Stratification: Artificial constancy. A number of points close to (hugging) the central line.

> *Rule 9:* Fifteen consecutive points within ± 1 sigma of the central line.

Mixture: A number of points hovering near the upper and lower 3 sigma limits with noticeable absence of points near the central line.

> *Rule 10:* Eight consecutive points outside ± 1 sigma of the central line.

Sudden jump in level: A sudden jump in level is characterized by a change in one direction. A number of points then appear on one side of the central line only.

> *Rule 11:* After a jump of ± 3 sigma, 3 consecutive points within ± 0.75 sigma of the jump point.

can be calculated from the actual data used to build the chart or can be calculated from *official* values of mean and/or sigma. The official values are usually obtained from a data set that has been chosen as the standard that is used to judge future operation.

Consider a variable collected by the real-time database with on-line electronic storage spanning weeks, months, or longer. To obtain the official values, the chart is configured and called for display, which also provides the calculated mean and sigma values and other chart parameters.

The chart display and calculations are updated by moving both forward and backward in time through the real-time database. This way one obtains the standard set of data and chooses the official values, which are then used in future process operation. Of course, the official values can be changed any time in the future as process operation analysis warrants.

Official values are not necessarily specification values. The official values are obtained from process data, which means that the process is capable of making product that can meet the requirements as specified by the official values. Thus official values are internal to the process/plant. Specification values, on the other hand, are externally imposed by customer requirements, which the process may or may not be able to meet. If the process cannot make product according to customer specifications, then it needs to be modified to do so.

Using official and calculated values for the mean and sigma, the rules of Table 2-3 can be evaluated four different ways. In addition, the upper and lower control limits and the central line of a chart can be calculated four different ways, as follows:

1. Calculated mean and sigma
2. Calculated mean and official sigma
3. Official mean and calculated sigma
4. Official mean and sigma

The rules are usually evaluated using the calculated mean and sigma.

The calculated mean is used to track the actual mean value of a process variable as it drifts during operation and if it is not essential to maintain the mean at a single value. Otherwise, the official mean is used.

The official sigma can be used when the variable is under closed-loop control. Consider real-time samples for a given variable collected over a period of time when feed flow and feed composition changes, as well as process conditions, were of a particular magnitude. Assume that the variation of the variable under these conditions is acceptable and it is not necessary to evaluate rules and alarm the operator. It is, indeed, desirable to take action when the upsets worsen and the variation increases. This is accomplished by choosing an official sigma and a set of rules that is violated only when the variation exceeds the acceptable variation. For example, the official sigma is chosen to have a value so that, when rules 1 and 3 of Table 2-3 are selected to indicate the in-statistical-control state of the variable, they are violated only when the upsets worsen and the variation increases above what is acceptable.

Chapter 3

Statistics
and Subgrouped Samples

This chapter presents the necessary statistical calculations for subgrouped samples. Moreover, it provides the calculations for mean and sigma values for subgroup means, ranges, and sigmas. Reference 4 covers some of this material in more detail. Other topics discussed are different ways to form subgroups and the rules to use for the xbar, range, and sigma values of a variable.

3.1 SUBGROUPING OF DATA

Samples of a given variable are subgrouped in such a way that samples within a subgroup are as much alike or homogeneous as possible and that subgroups differ from one another as much as possible. This is particularly difficult for continuous process variables where consecutive samples are correlated to one another. However, data can be subgrouped in different ways to achieve different objectives. Consider the following two ways of forming subgroups:

1. Size N: divide a group of samples into consecutive subgroups of size N.
2. Size N skip M: choose N consecutive samples for the subgroup and then skip M consecutive samples

Subgrouping method 1 is the typical way of subgrouping data. However, subgrouping method 2 is important to use to minimize the autocorrelation of contiguous samples. This makes the application of SPC charts more effective for real-time monitoring and it is discussed in Chapter 4.

3.2 XBAR, RANGE, AND SIGMA AND THEIR MEAN VALUES

A group of samples can be divided into a number of subgroups of equal size. This section shows the calculations for their mean, range, and sigma and their mean values.

The subgroup mean or xbar is

$$XB(j) = \frac{\sum_{i=1}^{N} X(i, j)}{N}, \qquad j = 1, \ldots, NS \tag{3-1}$$

The subgroup range is calculated by

$$R(j) = XMAX(j) - XMIN(j), \qquad j = 1, \ldots, NS \tag{3-2}$$

The subgroup sigma is computed by

$$S(j) = \sqrt{\frac{\sum_{i=1}^{N} [X(i, j) - XB(j)]^2}{N-1}}, \qquad j = 1, \ldots, NS \tag{3-3}$$

The xbar mean is computed by

$$XBB = \frac{\sum_{j=1}^{NS} XB(j)}{NS} \tag{3-4}$$

The range mean is computed by

$$RB = \frac{\sum_{j=1}^{NS} R(j)}{NS} \tag{3-5}$$

The sigma mean is computed by

$$SB = \frac{\sum_{j=1}^{NS} S(j)}{NS} \tag{3-6}$$

In Eqs. (3-1) through (3-6),

$$NS = \text{number of subgroups}$$
$$N = \text{subgroup size}$$
$$X(i, j) = i\text{th value of the variable in subgroup } j$$
$$XB(j) = j\text{th subgroup mean}$$
$$R(j) = j\text{th subgroup range}$$
$$XMAX(j) = j\text{th subgroup maximum value}$$
$$XMIN(j) = j\text{th subgroup minimum value}$$
$$S(j) = j\text{th subgroup sigma}$$
$$XBB = \text{Xbar mean}$$
$$RB = \text{range mean}$$
$$SB = \text{sigma mean}$$

3.3 CALCULATION OF XBAR AND RANGE SIGMA USING RANGE MEAN

Sigma prime is computed using the range mean as follows:

$$SIGP = \frac{RB}{d2} \tag{3-7}$$

Then the xbar sigma and range sigma are computed by

$$SIGXB = \frac{SIGP}{\sqrt{N}} \tag{3-8}$$

$$SIGR = \left(\frac{RB}{3}\right)(D4 - 1) \tag{3-9}$$

where, in Eqs. (3-7) through (3-9),

$$SIGXB = \text{xbar sigma}$$
$$d2 = \text{value given in Table C-1 in Appendix C}$$
$$SIGR = \text{range sigma}$$
$$D4 = \text{value given in Table C-2 in Appendix C}$$

3.4 CALCULATION OF XBAR AND SIGMA SIGMA USING SIGMA MEAN

Another way to compute sigma prime is by using the sigma mean as follows:

$$SIGP = \frac{SB}{c2} \sqrt{\frac{N-1}{N}} \tag{3-10}$$

Then the xbar sigma and sigma sigma are computed by

$$SIGXB = \frac{SIGP}{\sqrt{N}} \tag{3-11}$$

$$SIGS = \frac{SB}{3}(B4 - 1) \tag{3-12}$$

where, in Eqs. (3-10) through (3-12),

$c2$ = value given in Table C-1 in Appendix C
$B4$ = value given in Table C-3 in Appendix C
$SIGS$ = sigma sigma

3.5 WHEN TO USE SUBGROUP RANGES OR SIGMAS

As the subgroup size increases, the sigma is more accurate than the range. It is not advisable to use two types of charts, however, because it creates confusion for the operator. With the computations being done automatically, the ease of doing calculations using the range is no longer an asset. Therefore, the sigma values and their associated chart are recommended.

3.6 XBAR HISTOGRAM

The xbar histogram is needed for two reasons. One is to check the normality of the xbar values. The other is for process capability analysis, which is discussed in detail in Chapter 8.

The xbar histogram is built the same way as the individuals histogram presented in Section 2.3. It uses xbar values computed by Eq. (3-1) instead of individual samples. Also, referring to the normal distribution Eq. (2-4), XB is replaced by the xbar mean XBB computed by Eq. (3-4), and $SIGP$ is replaced by the xbar sigma $SIGXB$ computed by Eq. (3-8) or Eq. (3-11).

An analyst/operator should be able to configure and display xbar histograms using any subgrouping method and size and also be able to move through the real-time database and redisplay the histogram with the normal curve and recalculate skewness and kurtosis and other parameters.

The xbar values are accepted as normal when their skewness and kurtosis satisfy the relationships described by Eqs. (2-8) and (2-9), respectively.

3.7 STATISTICAL CONTROL RULES FOR XBAR, RANGE, AND SIGMA

The 11 rules of Table 2-3 are also used to check whether the calculated xbar values of a variable are in statistical control as they were for individual samples. In this case, the sigma used is the calculated xbar sigma $SIGXB$, and the central line used is the calculated xbar mean XBB.

It would be nice if the same set of rules could be used to check the statistical control state of the range values. Theoretically, they are not applicable, because the distribution of the ranges is not necessarily normal even though that of the xbars is. Practically, however, according to reference 5, for subgroup size greater than or equal to 4 or 5 the rules of Table 2-3 can also be applied to the ranges. By inference, then, the same rules can also be used for sigmas provided the subgroup size satisfies the above requirement. More information and analysis about which rules to use and why are provided later when specific types of charts and their applications are discussed.

Chapter 4

SPC and Process Dynamics and Minimum Autocorrelation

Continuous processes are characterized by dynamics, which vary depending on the unit, unit throughput, and other variables. Therefore, the dynamic response of a given variable needs to be considered when using SPC.

Another aspect of continuously sampled variables from dynamic processes is that contiguous samples tend to be correlated to one another. This chapter shows how to determine their autocorrelation and the subgrouping method to use to minimize it. Plotting charts with minimum autocorrelation makes them, indeed, much more useful, because then the rules can be applied with more confidence in indicating the not-in-statistical-control state of a variable.

In addition, the chapter also covers how to determine the cross-correlation of variables, which is useful in determining cause and effect.

4.1 SUBGROUP FORMATION BASED ON PROCESS DYNAMICS

Selecting subgroup size for continuous variables is based on their dynamic response. Figure 4-1 shows a typical dynamic response for a process variable. A step change in the manipulated variable results in a controlled variable response that can be approximated by a steady-state gain deadtime and lag. *Steady-state*

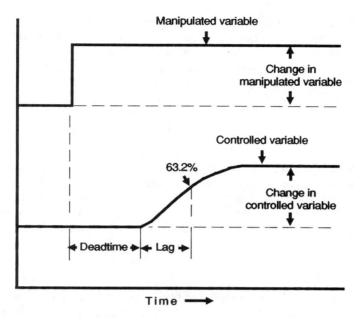

Figure 4-1 Typical Dynamic Response.

gain is defined as the final change in the controlled variable divided by the change in the manipulated variable. *Deadtime* is defined as the time it takes the controlled variable to start responding (0+) from the time the manipulated variable change was made. *Lag* is defined as the time it takes for the controlled variable to go from 0+% to 63.2% of its final value. The deadtime and lag represent the dynamic part of the response.

Whenever a causal variable changes due to a load upset, the traditional process controller responds to minimize in a desired fashion the transient or dynamic response of the controlled variable and drive it toward steady state. The time it takes to achieve this depends on the type and frequency of the causal variable changes, the process loop dynamics, and the values of the controller tuning parameters. The deadtime and lag determine the period of oscillation of the loop and the dynamic settings of controllers, for example, the integral and derivative settings of PID (proportional + integral + derivative) controllers.

In applying SPC it is advisable to consider process dynamics. The deadtime and lag or the period of oscillation are used to estimate how long it takes to reach steady state. A good estimate is

$$TSS = 3T0 \tag{4-1}$$

where

TSS = time to reach steady state
$T0$ = period of oscillation

The period of oscillation can be obtained from the values of the deadtime and lag or the controller tuning parameters or while the control loop is being tuned. Of course, an estimate of it is all that is necessary.

The subgroup should be chosen so that its elapsed time is TSS. If the sample period, the time between successive samples of a continuous variable is T, then the subgroup size is

$$N = \frac{3T0}{T} \tag{4-2}$$

Consider a temperature loop with $T0 = 5$ min and $T = 1$ min; substituting these values into Eq. (4-2) results in a subgroup size $N = 15$.

Real-time monitoring and alarming, however, usually require faster response from the SPC control chart. Then a reasonable compromise is to chose the subgroup so that its elapsed time is the period of oscillation; that is,

$$N = \frac{T0}{T} \tag{4-3}$$

In this case, the subgroup size for the preceding example is $N = 5$.

Using the second subgrouping method, increases the difference between subgroups, and thus minimizes their correlation, by increasing the skip size. On the other hand, waiting too long between subgroups may delay the effective implementation of control action. This is the compromise that needs to be made in the real-time monitoring and alarming application of SPC to continuous variables. The next two sections provide a practical approach to solving this problem.

4.2 AUTOCORRELATION AND CROSS-CORRELATION WITH TIME SHIFT

Consider an effect variable Y and a cause variable X. A question often asked concerns the effect of X on Y. For example, what is the effect of temperature on composition? The *cross-correlation coefficient* provides a quantitative measure of the correlation between two variables. Of course, this assumes that any other variables that cause Y to change when they change must be kept constant for the time span considered. If two cause variables change, then their ratio can be used, for example, the cross-correlation of composition Y versus the catalyst to feed ratio X or the ratio of two compositions Y versus the catalyst to feed ratio X.

In virtually all processes, particularly fluid processes, there is a certain time delay between a change in the cause variable and the corresponding change in the effect variable. For zero time delay, the values for each pair of Y and X samples must correspond as closely as possible to the same time. For time delay TD, there must be a shift of TD in the times for each sample pair, with the effect variable lagging the cause variable.

Consider the sample pairs $X(1)$, $Y(1 + TD)$, ..., $X(NS)$, $Y(NS + TD)$, where TD represents the time delay. The cross-correlation coefficient with time delay is computed as follows:

$$SX = \sum_{i=1}^{NS} X(i) \qquad (4\text{-}4)$$

$$SY(TD) = \sum_{i=1}^{NS} Y(i + TD) \qquad (4\text{-}5)$$

Then the mean values of X and Y are

$$XB = \frac{SX}{NS} \qquad (4\text{-}6)$$

$$YB = \frac{SY(TD)}{NS} \qquad (4\text{-}7)$$

Let

$$A = \sum_{i=1}^{NS} [Y(i + TD) - YB][X(i) - XB] \qquad (4\text{-}8)$$

$$B = \sum_{i=1}^{NS} [Y(i + TD) - YB]^2 \qquad (4\text{-}9)$$

$$C = \sum_{i=1}^{NS} [X(i) - XB]^2 \qquad (4\text{-}10)$$

Then the cross-correlation coefficient with time delay TD is

$$\text{rho}(TD) = \frac{A}{\sqrt{BC}} \qquad (4\text{-}11)$$

Least squares regression is used to fit the data to the straight line $Y = aX + b$. The coefficients a and b are computed as follows. Let

$$SYX = \sum_{i=1}^{NS} Y(i + TD)X(i) \qquad (4\text{-}12)$$

$$SXX = \sum_{i=1}^{NS} [X(i)]^2 \qquad (4\text{-}13)$$

Then the coefficients a and b are

$$a = \frac{NS\ SYX - SY\ SX}{NS\ SXX - SX^2} \tag{4-14}$$

$$b = \frac{SY\ SXX - SX\ SYX}{NS\ SXX - SX^2} \tag{4-15}$$

The plot of an effect variable versus a cause variable is referred to as the *scatter diagram* and is presented in Chapter 8. The regression line is usually superimposed on this diagram.

The cross-correlation coefficient ranges in value from -1.0 to $+1.0$. Zero means that the two variables do not affect one another. A value of $+1.0$ means that there is a strong positive correlation between them. That is, when one increases, the other increases also. A value of -1.0, on the other hand, means that there is a strong negative correlation between them. That is, when one increases, the other decreases. Values between 0 to 1.0 and 0 to -1.0 indicate that there may be positive or negative correlation, with confidence increasing as the values approach 1.0 or -1.0.

Now consider using the same variable for the effect and the cause. In this case, the cross-correlation coefficient rho(TD) of Eq. (4-8) becomes the autocorrelation coefficient with time shift TD. The *autocorrelation coefficient* is a quantitative measure of the correlation of a variable against itself for a given time delay. It ranges in value from $+1.0$ to -1.0.

4.3 USING SUBGROUP TYPE *N* SKIP *M* TO MINIMIZE AUTOCORRELATION

When the time delay is zero, the autocorrelation coefficient is 1.0, meaning that the variable is correlated to itself 100%, or maximum correlation. As the time delay increases, the autocorrelation decreases toward zero, and sometimes it continues in the negative direction, as shown in Fig. 4-2. Often it settles to a minimum value above zero. The time delay in Fig. 4-2 is plotted as a function of the period of oscillation, *T0*. Reference 6 provides more details for some of the calculations here. The same reference also recommends that the number of samples for calculating autocorrelation be greater than 50 to obtain a reliable estimate.

Samples of a continuous variable are correlated to one another, with the correlation decreasing as the time between samples increases until it reaches some minimum value. Thus, using Eq. (4-11) and the same variables for both X and Y, one can obtain the autocorrelation of a variable for different time delays.

Figure 4-2 shows a typical autocorrelation curve, with the autocorrelation being 1.0 when $TD = 0$. The objective is to choose a subgroup size N skip M with the skip size M chosen to obtain the minimum positive autocorrelation. This

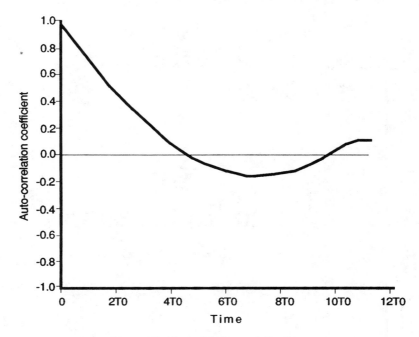

Figure 4-2   Typical Autocorrelation Curve.

is done by selecting the skip size M so that the total elapsed time for M is less than the time at which the autocorrelation function crosses the zero value for the first time.

In the case where the autocorrelation function settles above zero, the total elapsed time for M for minimum autocorrelation can be chosen as the time at which the absolute value of the slope, that is, the absolute value of the change in autocorrelation per period of oscillation $T0$, is less than a chosen minimum, for example, less than 0.05.

Chapter 5

Xbar
and Range/Sigma
Charts

Figure 5-1 shows the xbar and range chart, which is a double chart. The upper chart is the xbar and it is a plot of subgroup mean values together with their mean, the xbar mean, and the upper and lower control limits. This chart indicates the central tendency of the variable. The lower chart is a plot of subgroup range values together with the range mean and the range upper and lower control limits. This chart indicates the dispersion of the variable. The two charts are used together so that the central tendency and dispersion are simultaneously visible and are used in a complementary fashion for process operation.

Dispersion can also be obtained using the subgroup sigma values. Figure 5-2 shows the xbar and sigma chart, which is also a double chart, and it is similar to the xbar and range chart. The sigma chart is more accurate than the range chart, particularly for subgroup sizes greater than 10.

5.1 CONTROL LIMITS FOR XBAR, RANGE AND SIGMA CHARTS

Sections 3.3 and 3.4 show the calculations for xbars, ranges, xbar mean and sigma, and range mean and sigma.

The most commonly used control limits are the 3-sigma, which for normal

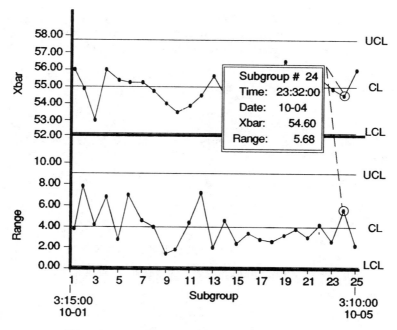

Figure 5-1   Xbar and Range Chart Using Table 5.1 Data.

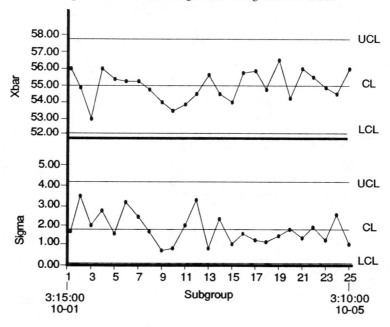

Figure 5-2   Xbar and Sigma Chart Using Table 5.1 Data.

distributions means that 99.74% of the values fall within three sigmas above and below the mean. However, there are situations where other limits are used.

The central line and control limits for the xbar chart are computed as follows:

$$CLXB = XBB \tag{5-1}$$

$$UCLXB = XBB + K1 \, SIGXB \tag{5-2}$$

$$LCLXB = XBB - K1 \, SIGXB \tag{5-3}$$

The central line and control limits for the range chart are computed as follows:

$$CLR = RB \tag{5-4}$$

$$UCLR = RB + K1 \, SIGR \tag{5-5}$$

$$LCLR = RB - K1 \, SIGR \tag{5-6}$$

$$\text{If } LCLR < 0, \text{ set } LCLR = 0 \tag{5-7}$$

The central line and control limits for the sigma chart are computed as follows:

$$CLS = SB \tag{5-8}$$

$$UCLS = SB + K1 \, SIGS \tag{5-9}$$

$$LCLS = SB - K1 \, SIGS \tag{5-10}$$

$$\text{If } LCLS < 0, \text{ set } LCLS = 0 \tag{5-11}$$

In Eqs. (5-1) through (5-11),

$K1$ = parameter for $K1$-sigma limits with default value 3.0, because 3.0 is the most commonly used control limit

$CLXB$ = central line xbar

$UCLXB$ = upper control limit xbar

$LCLXB$ = lower control limit xbar

CLR = central line range

$UCLR$ = upper control limit range

$LCLR$ = lower control limit range

CLS = central line sigma

$UCLS$ = upper control limit sigma

$LCLS$ = lower control limit sigma

The variable XBB is computed by Eq. (3-4), and $SIGXB$ is computed by Eq. (3-8) or Eq. (3-11), RB is computed by Eq. (3-5), $SIGR$ is computed by Eq. (3-9), SB is computed by Eq. (3-6), and $SIGS$ is computed by Eq. (3-12).

Let

$OXBB$ = official xbar mean

ORB = official range mean
OSB = official sigma mean

Then the official xbar sigma, $OSIGXB$, is computed either using ORB in Eq. (3-7) and then Eq. (3-8) or using OSB in Eq. (3-10) and then Eq. (3-11). The official range sigma, $OSIGR$, is computed using ORB in Eq. (3-9). The official sigma sigma, $OSIGS$, is computed using OSB in Eq. (3-12).

Section 4.6 showed that there are four combinations of rule evaluations and control limits for each chart that can be used based on selecting calculated and/or official values for mean and sigma.

5.2 EXAMPLE XBAR AND RANGE/SIGMA CHARTS

This section provides a specific numerical example for the xbar and range and for the xbar and sigma charts using the individual samples of Table 2-1. Table 5-1 lists the samples of Table 2-1 arranged in subgroups of four and the calculated values for xbar, range, and sigma for each subgroup. Figure 5-1 shows the resulting xbar and range chart, while Fig. 5-2 shows the resulting xbar and sigma chart.

Other parameters and calculations for the xbar and range chart are as follows:

$$N = 4 \qquad NS = 25 \qquad K1 = 3.0 \qquad XBB = 54.99$$
$$RB = 3.97 \quad SIGP = 1.93 \quad SIGR = 1.69 \quad SIGXB = 0.96$$
$$UCLXB = 57.87 \quad CLXB = 54.99 \quad LCLXB = 52.11$$
$$UCLR = 9.04 \quad CLR = 3.97 \quad LCLR = -1.10 \quad SETLCLR = 0.0$$

Other parameters and calculations for the xbar and sigma chart are as follows.

$$N = 4 \qquad NS = 25 \qquad K1 = 3.0 \qquad XBB = 54.99$$
$$SB = 1.78 \quad SIGP = 1.93 \quad SIGS = 0.75 \quad SIGXB = 0.96$$
$$UCLXB = 57.77 \quad CLXB = 54.99 \quad LCLXB = 52.11$$
$$UCLS = 4.03 \quad CLS = 1.78 \quad LCLS = -0.47 \quad SETLCLS = 0.0$$

Figure 2-2 shows the histogram for the individual samples. The skewness is -0.33 and the kurtosis is -0.03. Visual inspection of the histogram and the skewness and kurtosis values demonstrates that the distribution of the data is normal according to the rule given by Eqs. (2-8) and Eq. (2-9).

All 11 rules of Table 2-3 were checked for all charts, that is, the xbar and range and the xbar and sigma, and none was violated.

TABLE 5-1 XBAR, RANGE, AND SIGMA WITH SUBGROUP SIZE $N = 4$

Sample	Subgroup	Date	Time	Value	Xbar	Range	Sigma
1		10/01	00:05:00	54.57			
2		10/01	01:10:00	58.47			
3		10/01	02:06:00	54.96			
4	1	10/01	03:15:00	56.17	56.04	3.90	1.76
5		10/01	04:10:00	57.57			
6		10/01	05:08:00	49.47			
7		10/01	06:07:00	55.39			
8	2	10/01	07:12:00	56.39	54.78	7.80	3.46
9		10/01	08:10:00	52.12			
10		10/01	09:05:00	54.84			
11		10/01	10:18:00	50.64			
12	3	10/01	11:20:00	54.48	53.02	4.20	1.99
13		10/01	12:14:00	56.74			
14		10/01	13:15:00	52.48			
15		10/01	14:22:00	59.14			
16	4	10/01	15:25:00	55.50	55.96	6.66	2.77
17		10/01	16:18:00	56.18			
18		10/01	17:15:00	54.29			
19		10/01	18:12:00	54.15			
20	5	10/01	19;14:00	57.07	55.42	2.92	1.44
21		10/01	20:20:00	56.71			
22		10/01	21:17:00	58.78			
23		10/01	22:30:00	54.00			
24	6	10/01	23:35:00	51.68	55.29	7.10	3.10
25		10/02	00:18:00	53.75			
26		10/02	01:25:00	52.93			
27		10/02	02:06:00	57.56			
28	7	10/02	03:12:00	56.97	55.30	4.63	2.30
29		10/02	04:11:00	52.67			
30		10/02	05:08:00	56.73			
31		10/02	06:08:00	55.34			
32	8	10/02	07:12:00	54.42	54.79	4.06	1.70
33		10/02	08:15:00	53.07			
34		10/02	09:05:00	54.52			
35		10/02	10:17:00	54.52			
36	9	10/02	11:20:00	53.65	53.94	1.45	0.71
37		10/02	12:04:00	54.62			
38		10/02	13:15:00	52.99			
39		10/02	14:12:00	52.80			
40	10	10/02	15:20:00	53.38	53.45	1.82	0.82
41		10/02	16:18:00	53.12			
42		10/02	17:19:00	56.08			
43		10/02	18:12:00	54.76			
44	11	10/02	19:19:00	51.65	53.90	4.43	1.93
45		10/02	20:11:00	57.23			
46		10/02	21:17:00	56.17			
47		10/02	22:21:00	54.42			
48	12	10/02	23:32:00	50.01	54.46	7.22	3.18
49		10/03	00:13:00	54.56			
50		10/03	01:10:00	55.50			

(*continued*)

TABLE 5-1   XBAR, RANGE, AND SIGMA WITH SUBGROUP SIZE N = 4 (*Continued*)

Sample	Subgroup	Date	Time	Value	Xbar	Range	Sigma
51		10/03	02:06:00	55.87			
52	13	10/03	03:15:00	56.55	55.62	1.99	0.83
53		10/03	04:10:00	57.22			
54		10/03	05:08:00	55.23			
55		10/03	06:07:00	52.66			
56	14	10/03	07:12:00	52.66	54.44	4.56	2.21
57		10/03	08:10:00	55.09			
58		10/03	09:05:00	54.04			
59		10/03	10:18:00	52.56			
60	15	10/03	11:20:00	54.24	53.98	2.53	1.05
61		10/03	12:14:00	54.78			
62		10/03	13:15:00	54.34			
63		10/03	14:22:00	57.88			
64	16	10/03	15:25:00	56.22	55.80	3.54	1.60
65		10/03	16:18:00	57.37			
66		10/03	17:15:00	56.31			
67		10/03	18:12:00	54.48			
68	17	10/03	19:14:00	55.37	55.88	2.89	1.24
69		10/03	20:20:00	56.23			
70		10/03	21:17:00	54.57			
71		10/03	22:30:00	54.66			
72	18	10/03	23:35:00	53.46	54.73	2.77	1.14
73		10/04	00:10:00	55.27			
74		10/04	01:05:00	57.05			
75		10/04	02:06:00	58.38			
76	19	10/04	03:12:00	55.50	56.55	3.11	1.45
77		10/04	04:11:00	55.50			
78		10/04	05:08:00	54.76			
79		10/04	06:08:00	51.63			
80	20	10/04	07:12:00	55.02	54.23	3.87	1.76
81		10/04	08:15:00	57.32			
82		10/04	09:05:00	55.80			
83		10/04	10:17:00	56.60			
84	21	10/04	11:20:00	54.32	56.01	3.00	1.29
85		10/04	12:04:00	57.17			
86		10/04	13:15:00	53.02			
87		10/04	14:12:00	55.19			
88	22	10/04	15:20:00	56.86	55.56	4.15	1.90
89		10/04	16:18:00	56.00			
90		10/04	17:19:00	53.42			
91		10/04	18:12:00	55.78			
92	23	10/04	19:19:00	54.49	54.92	2.58	1.20
93		10/04	20:11:00	52.00			
94		10/04	21:17:00	53.18			
95		10/04	22:21:00	57.68			
96	24	10/04	23:32:00	55.56	54.60	5.68	2.53
97		10/05	00:02:00	56.76			
98		10/05	01:23:00	56.31			
99		10/05	02:06:00	56.31			
100	25	10/05	03:10:00	54.42	55.95	2.34	1.04

5.3 EXAMPLE OF USING *N* SKIP *M* TO MINIMIZE AUTOCORRELATION

This section presents an experiment using the xbar and sigma chart that demonstrates that using subgrouping method *N* skip *M* reduces the autocorrelation of data considerably, even for relatively small skip size *M* in relation to the subgroup size *N*.

Plotting charts with as little autocorrelation as possible renders them much more practical to use. This is because the chosen rules for determining the not-in-statistical-control state of a process variable can be applied with increased confidence.

Figure 5-3 shows a dynamic process simulation with a proportional plus integral (PI) controller. The process consists of a deadtime of 0.4 min and a lag of 0.4 min. The controller was tuned to provide the underdamped response of Fig. 5-4 with a period of oscillation, the time between successive peaks on the same side, of approximately $T0 = \frac{5}{3}$ min. A normal random number generator provides random values that are added to the controlled variable as shown.

The controlled variable with the added random numbers was sampled every minute. Thus, the sample period is $T = 1$ min. The subgroup size computed using Eq. (4-2) and the above values for TO and T is $N = 5$.

Using the above process with the dynamics as described, the following real-time experiment was done. The controller of Fig. 5-3 was placed on manual mode, and its output was changed and then placed on automatic mode to create several systematic load upsets to the simulated process. Then 250 samples were collected with a sample period of 1 min. The samples are listed in Table 5-2. The same samples are also plotted in Fig. 5-5.

Table 5-3 lists the auto-correlation coefficient, rho, for the samples of Table

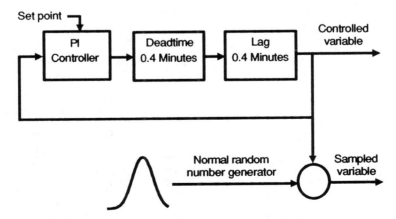

Figure 5-3   Simulated Process with Proportional Plus Integral (PI) Controller.

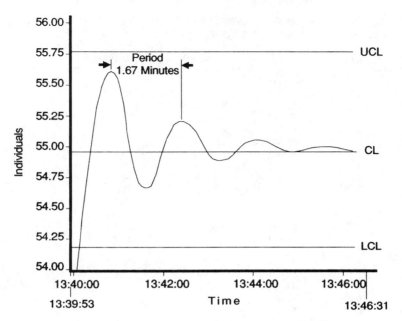

Figure 5-4   Closed-loop Response of Simulated Process with PI Controller.

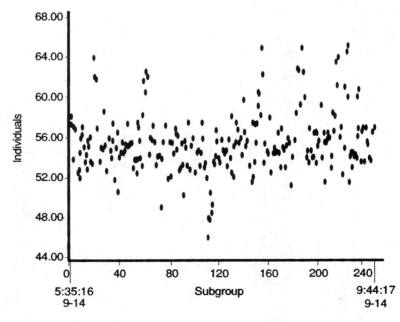

Figure 5-5   Plot of Table 5-2 Samples.

TABLE 5-2   ONE-MINUTE SAMPLES FROM THE SIMULATED PROCESS OF FIGURE 5-3 WITH MULTIPLE LOAD UPSETS

Sample	Value	Sample	Value	Sample	Value
1	57.445	51	52.974	101	54.589
2	58.225	52	53.885	102	54.811
3	53.819	53	53.962	103	53.257
4	57.059	54	52.420	104	57.460
5	56.772	55	53.962	105	54.524
6	52.681	56	57.513	106	56.010
7	52.865	57	56.932	107	53.052
8	54.502	58	55.836	108	53.477
9	52.071	59	54.116	109	52.120
10	56.162	60	53.296	110	46.188
11	56.341	61	58.333	111	48.062
12	57.187	62	61.668	112	47.986
13	54.931	63	60.567	113	50.419
14	53.711	64	62.501	114	48.521
15	54.299	65	62.016	115	49.393
16	52.791	66	54.227	116	53.622
17	55.683	67	56.085	117	53.242
18	55.205	68	55.760	118	54.134
19	56.078	69	57.354	119	56.795
20	53.672	70	55.636	120	55.851
21	63.934	71	53.230	121	54.107
22	62.023	72	54.238	122	53.611
23	61.802	73	54.329	123	55.862
24	57.043	74	49.924	124	54.720
25	53.767	75	53.863	125	54.413
26	55.107	76	55.617	126	54.747
27	55.099	77	55.715	127	51.123
28	54.389	78	57.279	128	55.882
29	58.607	79	51.835	129	54.148
30	55.127	80	55.485	130	52.573
31	52.654	81	52.522	131	58.227
32	55.882	82	55.609	132	51.801
33	54.673	83	54.409	133	55.255
34	54.030	84	56.952	134	55.510
35	55.714	85	56.313	135	57.524
36	57.449	86	54.818	136	56.859
37	51.817	87	56.179	137	55.567
38	55.037	88	52.740	138	55.957
39	56.533	89	52.816	139	54.320
40	50.572	90	55.505	140	59.724
41	54.035	91	53.238	141	56.643
42	54.349	92	50.353	142	55.219
43	55.674	93	55.696	143	55.868
44	54.434	94	54.964	144	53.118
45	54.582	95	53.309	145	51.680
46	55.304	96	57.526	146	57.583
47	57.482	97	52.815	147	52.213
48	55.155	98	55.172	148	53.690
49	55.265	99	56.085	149	57.386
50	55.456	100	52.559	150	55.545

(continued)

TABLE 5-2   ONE-MINUTE SAMPLES FROM THE SIMULATED PROCESS OF FIGURE 5-3 WITH MULTIPLE LOAD UPSETS (*Continued*)

Sample	Value	Sample	Value
151	57.326	201	55.771
152	59.107	202	59.219
153	60.456	203	54.021
154	60.352	204	56.668
155	58.399	205	51.698
156	64.991	206	54.096
157	62.278	207	55.553
158	53.459	208	55.527
159	55.354	209	55.917
160	54.672	210	56.606
161	54.320	211	56.078
162	58.070	212	52.923
163	52.488	213	54.261
164	52.504	214	58.398
165	54.559	215	56.885
166	54.984	216	58.263
167	54.489	217	63.472
168	54.664	218	61.344
169	56.613	219	64.016
170	54.493	220	56.157
171	53.174	221	57.218
172	57.007	222	52.376
173	54.911	223	54.959
174	54.821	224	60.910
175	53.026	225	64.699
176	55.293	226	60.187
177	56.022	227	65.159
178	55.378	228	51.764
179	51.320	229	54.228
180	54.120	230	54.663
181	55.827	231	53.170
182	58.400	232	54.639
183	62.809	233	54.263
184	62.709	234	54.442
185	59.337	235	53.024
186	64.970	236	59.953
187	62.539	237	56.290
188	59.115	238	56.984
189	54.005	239	60.848
190	54.864	240	53.791
191	53.597	241	56.557
192	57.026	242	56.996
193	54.844	243	53.645
194	54.049	244	57.106
195	54.120	245	55.708
196	56.588	246	55.237
197	56.638	247	54.099
198	53.436	248	53.762
199	56.431	249	56.614
200	55.252	250	57.133

TABLE 5-3 AUTOCORRELATION FOR SAMPLES OF TABLE 5-2

Time delay (TD) (min)	Autocorrelation coefficient (rho)	Time delay (TD) (min)	Autocorrelation coefficient (rho)
0	1.0000	16	0.0484
1	0.4949	17	0.0934
2	0.3753	18	0.0466
3	0.2710	19	0.0037
4	0.0618	20	0.0603
5	0.0774	21	0.0153
6	0.0742	22	0.0116
7	0.0128	23	−0.0356
8	0.0108	24	−0.0489
9	−0.0126	25	−0.0140
10	−0.0102	26	−0.0237
11	−0.0213	27	−0.0003
12	−0.0248	28	−0.0012
13	−0.0082	29	0.0127
14	0.0159	30	0.0695
15	0.0855		

5-2 as a function of time delay, TD, in minutes. Based on the discussion in Section 4.3, its minimum positive value before it crosses zero the first time is 0.0108, for which the time delay is 8 min. As can be seen from the values of Table 5-2, the autocorrelation function is reduced dramatically even for a short time delay in relation to the dynamics of the simulated process with the controller.

Figure 5-6 shows the xbar and sigma chart for the samples of Table 5-2 with the subgroup size $N = 5$ and no skipping; that is, the skip size $M = 0$. Since the number of samples is 250, this results in a chart with 50 subgroups.

All 11 rules of Table 2-3 were checked for both the xbar and sigma charts. The charts indicate the subgroups that violate at least one rule by enlarging the size of their symbol, as shown. The following list shows which rules were violated by which subgroups in parentheses for both the xbar and sigma charts:

XBAR CHART

Rule 1: Freak (2, 13, 22, 23, 31, 37, 38, 44)
Rule 3: Grouping 2 of 3 (21–24, 30–33, 36–39, 43–46)
Rule 4: Grouping 4 of 5 (22–26)
Rule 6: Downshift (15–26)
Rule 11: Jump (33–36)

SIGMA CHART

Rule 1: Freak (32, 38, 45, 46)
Rule 3: Grouping 2 of 3 (44–47)

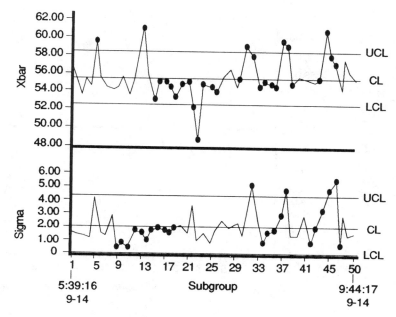

Figure 5-6 Xbar and Sigma Chart for the Samples of Table 5.2 with $N = 5$ and No Skipping.

Rule 6: Downshift (9–19)

Rule 7: Uptrend (34–38, 42–46)

The autocorrelation value is 0.0774 for a time delay of 5 min, which for all practical purposes minimizes the effect of autocorrelation. The skip size is chosen to cover the elapsed time of 5 min, and when divided by the sample period of 1 min, it results in a skip size of $M = 5$.

Thus using the subgrouping method of $N = 5$ and skip size $M = 5$ results in an xbar and sigma chart with 25 subgroups and uses all 250 samples of Table 5-2. Figure 5-7 shows the xbar and sigma chart for the samples of Table 5-2 with the subgroup size $N = 5$ skip size $M = 5$.

All 11 rules of Table 2-3 were checked for both the xbar and the sigma charts. The following list shows which rules wee violated by which subgroups in parentheses, both for the xbar and the sigma charts:

XBAR CHART

Rule 1: Freak (11, 19, 22)

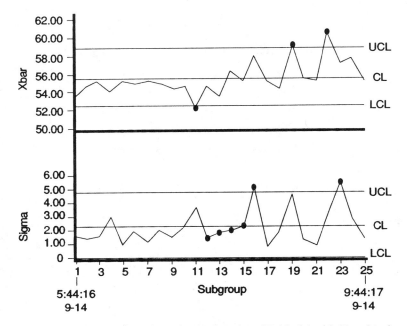

Figure 5-7   Xbar and Sigma Chart for the Samples of Table 5-2 with $N = 5$ and $M = 5$.

SIGMA CHART

Rule 1: Freak (16, 23)

Rule 7: Uptrend (12–16)

The results show clearly the dramatic reduction in rule violations when skip size $M = 5$, which reduces autocorrelation substantially, is used. Section 9.3 shows the effect of the same skip size for the CUSUM chart.

5.4 RULES TO USE WHEN THE VARIABLE IS NOT CONTROLLED

In every process and manufacturing plant, many variables are not under closed-loop control for various reasons, such as lack of appropriate on-line sensors and long time to obtain analysis results. These variables are usually the results of laboratory analysis, and they are entered in the real-time database and, together with the controlled variables, are used to monitor process operation. The xbar and range/sigma chart is used for real-time monitoring of these type of variables.

Rule 9, stratification, should not be used when the objective is to keep the charted variable as close as possible to a desired target, even though the variable is *not* under closed-loop control. Rule 10, mixture, should be used when the variable represents the blend of a mixture of products that come from more than one unit. Rule 4 is similar to rule 3, and rule 3 is preferable.

Instead of using the xbar chart with rules 5 and 6, gradual upward/downward change in level, rules 7 and 8, upward/downward trend, and rule 11, sudden jump in level, the author recommends the use of the CUSUM chart presented in Chapter 9. In many cases the CUSUM chart is more effective and simpler to use for real-time monitoring and alarming.

The author recommends the use of the xbar range/sigma chart and the CUSUM chart to complement one another. It is also recommended that no more than two rules be used for the xbar and range/sigma charts. In particular, rules 1 and 3 are usually the recommended rules.

5.5 RULES TO USE FOR RANGE OF CONTROLLABILITY MONITORING

When the charted variable is under closed-loop control, the controller responds to load upsets and maintains the variable as close to the desired set point as possible. First consider xbar chart rule violations for the variable under closed-loop control.

Rule 9, stratification, should not be used when the charted variable is under closed-loop control, because if the controller does its job and the load upsets are not continuous, which is usually the case, the end result is stratification. Rule 10, mixture, does not apply in usual control situations, because mixtures come from more than one unit and control is typically on a per unit basis.

Rules 1, 2, and 3 are the three recommended rules to use for variables under closed-loop control.

When the variable is controlled, its set point is set by the operator, a program, or another controller in the case of cascaded control loops. In these situations, the calculated value of the mean should be used for rule evaluation and for the limits.

Consider Fig. 5-7, which shows the xbar and sigma for the simulated process with the multiple load upsets. Now assume that this is the acceptable range of controllability for the controller and for the underlying process and load upsets. Within this range it is desired that rules 1 and 3 not be violated, either by the xbar or the sigma charts. This is achieved by using the calculated mean and an official sigma mean value large enough so that rules 1 and 3 are not violated for the xbar and sigma chart of Fig. 5-7. Of course, variations larger than the acceptable range will violate the rules and the operator will be notified.

Another way to monitor range of controllability is to chart the error variable of the controller, which is the difference of the set point minus the measurement. In this case, the desired mean value of the error is zero. Therefore, an official mean value of zero should be used.

5.6 LONG-TERM TRENDING

A typical closed-loop control trend consists of the measurement or controlled variable, the set point, and the controller output or manipulated variable. Real-time monitoring alarms are based on high and low absolute deviation from set point and the rate of change of the controlled variable. In addition, there are absolute alarms and limits on the manipulated variable. These trends provide for fast decision making when it is necessary to maintain the process under control and operating safely.

The xbar and range/sigma charts, on the other hand, should also be used for on-demand trending of the controlled variable to provide operation in the longer term and avoid making operational changes prematurely. This is accomplished by virtue of subgrouping samples based on process dynamics to minimize correlation, as described in Sections 4.1 and 4.3. Thus the operator using these charts is forced to make decisions on a longer-term basis relative to the standard trends.

Chapter 6

Scatter Diagram
with Time Shift
for Autocorrelation
and Cross-correlation

The scatter diagram is a plot of one variable against another, thus providing the ability to see and compute the cross-correlation between two variables. A scatter diagram of a variable plotted against itself with time shift provides the ability to see and compute the autocorrelation of a variable.

6.1 SCATTER DIAGRAM AND CROSS-CORRELATION

Figure 6-1 shows a scatter diagram of one variable plotted against another for visual display of cross-correlation between two variables. The calculations for cross-correlation with time shift and the linear regression line coefficients given in Section 4.1 are combined with the display to increase its utility. The regression line itself is also plotted with the diagram. All other calculations and parameters are shown upon request via windows.

Table 6-1 provides sample values for X and Y and the resulting calculations without any time shift, that is $TD = 0$, using Eqs. (4-4) to (4-15).

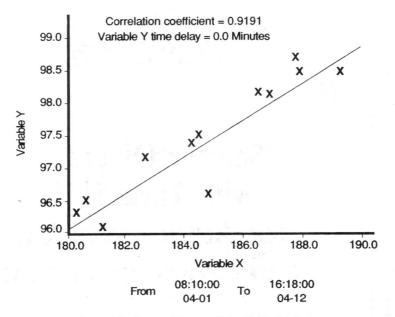

Figure 6-1   Scatter Diagram Using Table 6.1 Data.

The cross-correlation coefficient, rho, calculated using Eq. (4-11), and the regression line coefficients a and b, calculated using Eqs. (4-4) and (4-15), are

$$\text{rho}(0) = 0.9191, \qquad a = 0.276, \qquad b = 46.3616$$

Figure 6-1 shows the scatter diagram for this example.

To derive a meaningful correlation, the variables must be aligned in time. Thus, if one variable has fewer samples than the other over the same time span, the more plentiful samples are reduced in number to values corresponding to the times of the fewer samples.

As discussed in Section 4.1, the optional time delay is used to take into account the time delay between a change in the cause variable and the corresponding change in the effect variable. Time delay characterizes all dynamic processes, and it needs to be taken into account in correlation studies.

Figure 6-1 provides a visual display of the correlation between the two variables. The value of the cross-correlation coefficient is also computed and displayed to provide a quantitative measure of the correlation between the variables. A linear regression line is also computed and plotted on the diagram. The cross-correlation coefficient ranges in value from -1.0 to $+1.0$.

Positive values for the coefficient mean that as one variable increases the other increases also. Zero means that there is no correlation between the two variables, with correlation increasing as the coefficient increases in value up to

TABLE 6-1 SCATTER DIAGRAM SAMPLES AND CALCULATIONS

Subgroup	Date	Time	X	Y	$(X - XB)(Y - YB)$	$(Y - YB)^2$	$(X - XB)^2$	XY	X^2
1	04/01	08:10:00	187.8	98.7	3.7615	1.4803	9.5584	18,535.86	35,268.84
2	04/02	06:07:00	184.8	96.6	-0.0810	0.7803	0.0084	17,851.68	34,151.04
3	04/03	07:12:00	187.9	98.5	3.2449	1.0336	10.1867	18,508.15	35,306.41
4	04/04	08:10:00	180.5	96.6	3.7174	0.7803	17.7101	17,436.30	32,580.25
5	04/05	09:05:00	186.5	98.2	1.2840	0.5136	3.2101	18,314.30	34,782.25
6	04/06	10:18:00	181.2	96.1	4.8532	1.9136	12.3084	17,413.32	32,833.44
7	04/07	11:20:00	180.3	96.4	4.7757	1.1736	19.4334	17,380.92	32,508.09
8	04/08	12:14:00	182.7	97.2	0.5690	0.0803	4.0334	17,758.44	33,379.29
9	04/09	13:15:00	186.8	98.1	1.2899	0.3803	4.3751	18,325.08	34,894.24
10	04/10	14:22:00	184.5	97.5	-0.0035	0.0003	0.0434	17,988.75	34,040.25
11	04/11	15:25:00	184.3	97.4	0.0340	0.0069	0.1667	17,950.82	33,966.49
12	04/12	16:18:00	189.2	98.5	4.5665	1.0336	20.1751	18,636.20	35,796.64
			SX 2216.5	SY 1169.8	A 28.0117	B 9.1767	C 101.2092	SYX 21,6099.82	SXX 40,9507.23
			XB 184.708	YB 97.483					

a maximum of 1.0, that is, 100%, which indicates that there is maximum effect between the two variables.

Negative values for the coefficient mean that as one variable increases the other decreases. Zero means that there is no correlation between the two variables, with correlation increasing as the coefficient decreases in value up to a minimum of −1.0, that is, −100%, which indicates that there is maximum negative effect between the two variables.

The linear regression line is the optimum least squares fit to a straight line of the actual data. If one computed the difference between an actual value and the corresponding value obtained from the straight line, square it, and sum the error for all samples, the value of the summed error squared is minimum (least), and hence least squares.

6.2 SCATTER DIAGRAM AND AUTOCORRELATION

The scatter diagram can also plot a variable against itself, thus providing the ability to see and compute its autocorrelation. The time delay is then used to compute the autocorrelation of the variable for different time delays.

As mentioned in Section 4.1, the autocorrelation coefficient is a quantitative measure of the correlation of a variable against itself for a given time delay. Its value ranges from +1.0 to −1.0. When the time delay is zero, the autocorrelation coefficient is 1.0, meaning that the variable is correlated to itself 100%, or maximum correlation.

The scatter diagram is also used on demand to obtain the time delay required to obtain minimum autocorrelation as described in Section 4.3.

6.3 OTHER REQUIRED FUNCTIONS FOR SCATTER DIAGRAMS

So far the scatter diagram has been used to determine the cross-correlation between two variables and the autocorrelation of a single variable. There are times, however, when it is necessary to determine the cross-correlation of an effect variable with the ratio of two cause variables. An example is a continuous chemical reactor that uses a catalyst for its reaction. The ratio of catalyst to feed flow is one of the main variables that determines product composition. Therefore, the scatter diagram needs to be configured to access feed flow, catalyst flow, and product composition samples from the real-time database, compute the ratio of catalyst to feed, and to display the scatter diagram. This scatter diagram provides all calculations and features described in the previous two sections.

The feed and catalyst flows usually are sampled and collected periodically, say at a sample period of 1 to 5 min. The product composition may be also collected periodically from an on-line sensor or obtained from laboratory analysis and entered manually into the real-time database, say every hour or so. Thus the scatter diagram needs to be built with catalyst to feed ratio values that are closest to the sample dates/times of the manually entered composition values, taking into account the required time delay.

Another useful feature is the provision to collect two cause and two effect variables and build a scatter diagram for the ratio of the two effect variables versus the ratio of the two cause variables, for example, the ratio of two product composition variables versus the ratio of catalyst to feed flow for the chemical reactor.

Other mathematical transformations can be used to determine the cross-correlation of a function of an effect with a function of a cause variable. An example is the logarithm of composition versus temperature.

Chapter 7

Individuals Chart

Figure 7-1 shows the individuals chart. It is a plot of individual samples together with their mean and the upper and lower control limits. It is used to judge the variable's central tendency and dispersion as compared to the mean and sigma.

As in the case of the xbar and range charts, the rules and limits can be evaluated and computed four different ways using calculated and/or official values for the mean and sigma.

7.1 CONTROL LIMITS FOR INDIVIDUALS CHART

Equations (2-1) to (2-3) show the calculations for the mean XB, the sigma S, and the population sigma $SIGP$ for individual samples.

The central line and control limits are computed as follows:

$$CL = XB$$

$$UCL = XB + K1 \; SIGP$$

$$LCL = XB - K1 \; SIGP$$

Let OXB be the official value for the mean and let $OSIGP$ be the official value

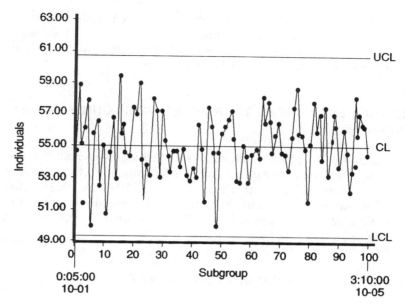

Figure 7-1   Individuals Chart Using Table 2.1 Data.

for the sigma. Then the central line and limits above and the rules can be evaluated four different ways using the desired combinations of *XB*, *SIGP*, *OXB*, and *OSIGP*, as described in Section 4.6.

7.2 EXAMPLE INDIVIDUALS CHART

The samples of Table 2-1 are used to build the individuals chart shown in Fig. 7-1. Note that the *x* axis shows the subgroup number as well as the time span of the real-time data. Other parameters and calculated values are as follows:

$NS = 100$	$K1 = 3.0$	$XB = 54.99$	$SIGP = 1.90$
$UCL = 60.69$	$CL = 54.99$	$LCL = 49.29$	

7.3 RULES TO USE AND REAL-TIME MONITORING

When the individuals chart is chosen for real-time monitoring instead of the xbar and range/sigma chart, the analysis, conclusions, and recommendations given in Sections 5.4 through 5.6 are directly applicable to this chart as well. Thus the individuals chart can be used for real-time monitoring of controlled variables to monitor range of controllability and can also be used for monitoring variables that are not under closed-loop control.

Chapter 17 presents the analysis that needs to be done to determine the applicability of the individuals chart.

7.4 INDIVIDUALS CHART FOR VARIABLES WITH LONG SAMPLE PERIODS

Using this chart for continuously sampled variables results in premature adjustment of set points and other variables to maintain statistical control, as compared to the xbar and range, xbar and sigma, and CUSUM charts. However, for variables that are infrequently sampled and for variables with sample periods, that is, time between samples, longer than an hour, the individuals chart may be the only one that is feasible for use.

7.5 TRENDING WITH THE INDIVIDUALS CHART

The individuals chart is similar to the trend displays used in traditional control, with the addition of the mean value, the upper and lower control limits, and evaluation of the rules.

Chapter 8

Process Capability Analysis

Sections 2.3 and 3.6 introduced the individuals and the xbar histogram, respectively. Section 2.4 provides the calculated values of skewness and kurtosis and their use as measures of the normality of a particular distribution of samples.

Another use of the histogram is for process capability analysis. Process capability is defined in relation to the upper and lower specification limits for a given variable. Thus the process is considered capable when the percentage of samples of a variable for that process that fall within the upper and lower specification limits is greater than a specified value. Another way of saying it is that the process is capable when the percentage of samples outside the specification limits is less than a specified value.

More precisely, process capability is determined by computing certain indexes based on a set of samples from the process.

8.1 PROCESS CAPABILITY USING THE INDIVIDUALS HISTOGRAM

The most commonly used indexes for process capability are the following:

$$\text{midpoint} = \frac{USL + LSL}{2} \tag{8-1}$$

$$\text{Tolerance} = USL - LSL \tag{8-2}$$

$$K = \frac{2(XB - \text{midpoint})}{\text{tolerance}} \tag{8-3}$$

$$CP = \frac{\text{tolerance}}{6\ SIGP} \tag{8-4}$$

$$CR = \frac{6\ SIGP}{\text{tolerance}} \tag{8-5}$$

$$CPU = \frac{USL - XB}{3\ SIGP} \tag{8-6}$$

$$CPL = \frac{XB - LSL}{3\ SIGP} \tag{8-7}$$

$$CPK = \min(CPU, CPL) \tag{8-8}$$

where, for Eqs. (8-1) through (8-8),

XB = mean computed using Eq. (2-1)
$SIGP$ = sigma prime computed using Eq. (2-3)
USL = upper specification limit
LSL = lower specification limit
K = process mean versus specification midpoint
CP = inherent process capability
CR = capability ratio
CPU = process capability based on the worst-case view of the data from USL
CPL = process capability based on the worst-case view of the data from LSL
CPK = process capability based on the worst-case view of the data

The value of K indicates the closeness of the process mean to the specification midpoint:

If $K > 0$, the process mean is above midpoint; $K < 0$ indicates that the process mean is below the midpoint.

If the upper and lower specification limits are symmetric, then a value of 1.0 or -1.0 indicates that the process mean is at the upper or lower limit, respectively.

When $K = 0$, the process mean is equal to the specification midpoint, and the value of CP or CR can be used to determine the process capability.

The meanings for CP and its inverse CR are the following:

$CP > 1.0$ or $CR < 1.0$ means that the process is capable; that is, the process variation is within the specified tolerance.

$CP < 1.0$ or $CR > 1.0$ means the process is not capable.

CPK is interpreted as follows:

$CPK < 0$ indicates that the process mean is outside the specification limits (USL, LSL).

$CPK = 0$ indicates that the process mean is equal to one of the specification limits.

$CPK > 0$ indicates that the process mean is within the specification limits.

$CPK = 1.0$ means that one side of the 6-sigma limits falls on a specification limit.

$CPK > 1.0$ means that the 6-sigma limits fall completely within the specification limits.

CP or CR may indicate that the process is capable, but the process mean can be way outside the specification limits. The value of K, on the other hand, shows where the mean lies. Therefore, CP or CR should be used together with K to ascertain that the process is capable and its mean value is within the specification limits. In particular, K close to zero indicates the process mean is close to the specification midpoint.

The $CPK > 1.0$, however, indicates that both the mean and the 6-sigma limits are within the specification limits.

The capability indexes are quantitative measures for process capability. However, the distribution of the samples should also be considered. In particular, the normality of samples should be determined before embarking on capability analysis. Section 2.3 described the use of the individuals histogram with the superimposed normal curve for visual inspection of normality. In addition, it provided the quantitative measures of skewness and kurtosis and the rule of thumb, Eqs. (2-8) and (2-9), that suggests when to accept the samples as normal.

Once the distribution of the samples is accepted as close enough to normal, then Eq. (2-4) is used to obtain the area of the normal curve above the USL and convert it to percent. This is the percent out of spec high and is designated as $ANHIGH$, the area of normal above the upper specification limit. Similarly, Eq. (2-4) is used to obtain the area of the normal curve below the LSL and convert it to percent. This is the percent out of spec low and is designated as ANLOW, the area of normal below the lower specification limit.

The total area of the normal curve off specification, $ANOFSPEC$, is

$$ANOFSPEC = ANHIGH + ANLOW \qquad (8\text{-}9)$$

Figure 8-1 shows the individuals histogram for the samples of Table 2-1 with the upper and lower specification limits and the target for completeness. The computed process capability indexes, off-specification area for the normal curve, skewness and kursosis, central line, upper and lower control limits, and other parameters are:

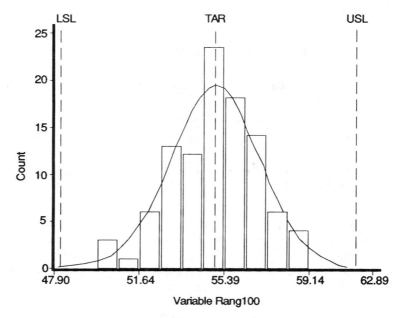

Figure 8-1   Process Capability Using Table 2.1 Data.

$NS = 100$	$XB = 54.99$	$SIGP = 1.90$
$UCL = 60.67$	$CL = 54.99$	$LCL = 49.30$
$USL = 62.00$	$TAR = 55.00$	$LSL = 48.00$
skewness $= -0.33$	kurtosis $= -0.03$	
$K = 0.00$	$CP = 1.23$	$CR = 0.81$
$CPU = 1.23$	$CPL = 1.23$	$CPK = 1.23$
$ANHIGH = 0.01\%$	$ANLOW = 0.01\%$	$ANOFSPEC = 0.02\%$

A histogram used for the analysis of real-time data should provide the display of Fig. 8-1 and all the calculated values above.

8.2 PROCESS CAPABILITY USING THE XBAR HISTOGRAM

The analysis of the previous section applies for individual samples that satisfy or can be made to satisfy, using mathematical transformations, the normality rule of thumb. What happens then if they do not?

When the samples come from dynamic processes and are correlated, then the procedures of Chapter 4 should be followed. In particular, subgroup formation should take into account the dynamics of the process, and subgroup type size N

skip M, with M large enough to minimize correlation, should be used to build an xbar histogram and perform the process capability analysis. Indubitably, real-time process capability analysis with the xbar histogram as described here is the effective way to do it, because of the many correlated dynamic variables that characterize process industry plants and automated manufacturing plants.

Chapter 9

Drift Detection Using the CUSUM Chart

The CUSUM chart described in references 7 and 8 is designed to detect sustained shifts of the mean value from a desired target. It computes a normalized cumulative deviation of the subgroup mean, minus the target, divided by the subgroup sigma. In general, it detects smaller changes in the mean more rapidly than the xbar chart. Because it is iterative, it lends itself to real-time monitoring.

The author has expanded the CUSUM scheme described in references 7 and 8 to make it more useful for nonlinear processes. This is achieved by providing different slack values and decision intervals for the accumulated normalized deviations above and below the target value.

In addition, the author has invented a nonlinear controller that uses the expanded CUSUM scheme calculations and provides closed-loop control. The controller is described in Chapter 10.

9.1 CALCULATIONS AND EXAMPLE CUSUM CHART

The desired group of individual samples for a given variable is obtained from the real-time database starting from the current time backward. These samples are then used to form subgroups of equal size by one of the subgrouping methods described in Section 3.2.

The scheme accumulates normalized deviations more than $k1$ units above or more than $k2$ units below the target TAR. The constants $k1$ and $k2$ serve as slack values. The normalized deviation from target is computed by

$$Y(j) = \frac{XB(j) - TAR}{S(j)}, \quad j = 1, \ldots, NS \qquad (9\text{-}1)$$

The cumulative sum high, that is, above the target, is computed as follows:

$$SH(j) = \max[0, SH(j - 1) + Y(j) - k1], \quad j = 1, \ldots, NS \qquad (9\text{-}2)$$

The cumulative sum low, that is, below the target, is computed as follows:

$$SL(j) = \max[0, SL(j - 1) - Y(j) - k2], \quad j = 1, \ldots, NS \qquad (9\text{-}3)$$

In Eqs. (9-1) through (9-3),

$XB(j) = j$th subgroup mean computed using Eq. (3-1)
$S(j) = j$th subgroup sigma computed using Eq. (3-3)
$TAR = $ target value
$Y(j) = j$th subgroup normalized deviation from target
$SH(j) = $ cumulative sum high, above target, up to and including the jth subgroup
$SL(j) = $ cumulative sum low, below target, up to and including the jth subgroup
$k1 = $ slack variable above target
$k2 = $ slack variable below target

Now let

$h1 = $ decision interval high, above target
$h2 = $ decision interval low, below target

The slack variables and the decision intervals must satisfy the following conditions:

$$k1 \geq 0 \quad \text{and} \quad k2 \geq 0 \qquad (9\text{-}4)$$

$$h1 \geq k1 \quad \text{and} \quad h2 \geq k2 \qquad (9\text{-}5)$$

SH is used to detect a departure from target on the high side, while SL is used to detect a departure from target on the low side. Both are used simultaneously in two-sided CUSUM monitoring schemes. When the value of SH exceeds the decision interval value $h1$, or the value of SL exceeds the decision interval $h2$, the process is not in statistical control.

The properties of the CUSUM scheme are determined by the decision intervals $h1$, $h2$, and the slack values $k1$, $k2$. They are the tuning parameters. When $k1 = k2 = k$ and $h1 = h2 = h$, this scheme reduces to that described in references 7 and 8. These references recommend the values $k = 0.5$ and $h = 4.0$ or 5.0 for detecting small sustained deviations in the mean.

SH and SL are initialized to zero for standard CUSUM schemes. The time

TABLE 9-1   CUSUM SUBGROUPS OF SIZE 4

Subgroup	Xbar	Sigma	Y	$Y - k1$	SH	$-Y - k2$	SL
1	56.04	1.76	1.159	0.659	0.659	−1.659	0.000
2	54.78	3.46	0.225	−0.275	0.384	−0.725	0.000
3	53.02	1.99	−0.492	−0.992	0.000	−0.008	0.000
4	55.96	2.77	0.707	0.207	0.207	−1.207	0.000
5	55.42	1.44	0.986	0.486	0.693	−1.486	0.000
6	55.29	3.10	0.416	−0.084	0.609	−0.916	0.000
7	55.30	2.30	0.565	0.065	0.674	−1.065	0.000
8	54.79	1.70	0.465	−0.035	0.639	−0.965	0.000
9	53.94	0.71	−0.084	−0.584	0.055	−0.416	0.000
10	53.45	0.82	−0.671	−1.171	0.000	0.171	0.171
11	53.90	1.93	−0.052	−0.552	0.000	−0.448	0.000
12	54.46	3.18	0.145	−0.355	0.000	−0.645	0.000
13	55.62	0.83	1.952	1.452	1.452	−2.452	0.000
14	54.44	2.21	0.199	−0.301	1.151	−0.699	0.000
15	53.98	1.05	−0.019	−0.519	0.632	−0.481	0.000
16	55.80	1.60	1.125	0.625	1.257	−1.625	0.000
17	55.88	1.24	1.516	1.016	2.273	−2.016	0.000
18	54.73	1.40	0.521	0.021	2.294	−1.021	0.000
19	56.55	1.45	1.758	1.258	3.552	−2.258	0.000
20	54.23	1.76	0.131	−0.369	3.183	−0.631	0.000
21	56.01	1.29	1.558	1.058	4.241	−2.058	0.000
22	55.56	1.90	0.821	0.321	4.562	−1.321	0.000
23	54.92	1.20	0.767	0.267	4.829	−1.267	0.000
24	54.60	2.53	0.237	−0.263	4.566	−0.737	0.000
25	55.95	1.04	1.875	1.375	5.941	−2.375	0.000

$TAR = 54$, $k1 = 0.5$, $k2 = 0.5$.

interval from one out-of-control alarm to the next is referred to as the run length RL.

Using the 100 samples and the computed values of xbar and sigma from Table 5-1, the calculated values for SH and SL are shown in Table 9-1. Other parameters and calculations are as follows:

$$N = 4 \qquad NS = 25 \qquad XBB = 54.99 \qquad SB = 1.78 \qquad SIGXB = 0.96$$
$$TAR = 54.0 \qquad k1 = 0.5 \qquad k2 = 0.5 \qquad h1 = 4.0 \qquad h2 = 4.0$$

Figure 9-1 shows the CUSUM chart for this example. The symbol Y indicates the value of the normalized deviation from target as computed by Eq. (9-1). The symbol S indicates either cumulative sum high, SH, or cumulative sum low, SL. Although the computed value of SL, Eq. (9-3), is always positive or zero, it is

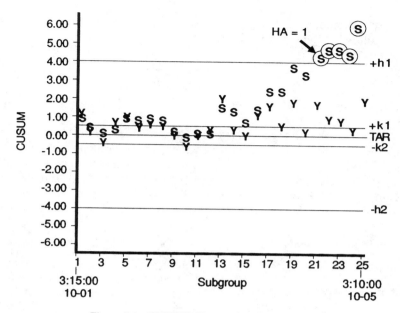

Figure 9-1   CUSUM Chart Using Table 9.1 Data.

plotted as a negative value to indicate that it is moving below and away from the desired target.

The normalized deviation Y, as well as the cumulative sum S, is mostly drifting away above the target, and it first exceeds the value of $h1 = 4.0$ at subgroup 21, at which point it is not in statistical control, indicated by setting the high alarm, HA = 1 (true). The variable continued drifting away because no operator action was taken at subgroup 21. This will be discussed further later in this chapter.

9.2 TUNING THE CUSUM SCHEME

The desired mean value for the variable is the target value at which the process should operate. Another value required for the design is the acceptable change in mean value from the target that the CUSUM scheme should detect.

References 7 and 8 discuss the symmetrical CUSUM scheme, in which case the slack values are equal to each other, and the decision intervals are also equal to each other. In this case, these references suggest that the slack value be chosen so that the average time between out-of-statistical-control signals is maximized, and the average time to signal an unacceptable shift in the mean is minimized.

They recommend that $k = 0.5$ and $h = 4.0$. These values are a special case of the following general relationships:

$$k = \frac{\text{acceptable change in mean}}{2\ SIGXB} \qquad (9\text{-}6)$$

The acceptable change in mean is in either direction, that is, above or below the target.

When the change in mean is chosen to be equal to $SIGXB$, the above relationships give $k = 0.5$, which is the value recommended by references 7 and 8. The same references suggest using $h = 4.0$ or 5.0 when $k = 0.5$. The value $h = 4.0$ is chosen for the remainder of the discussion. Based on the recommended values then, a useful rule of thumb is

$$h + k = 4.0 + 0.5 = 4.5 \qquad (9\text{-}7)$$

Solving for h yields

$$h = 4.5 - k \qquad (9\text{-}8)$$

And k and h must satisfy the conditions

$$k \geq 0 \qquad (9\text{-}9)$$

$$h \geq k \qquad (9\text{-}10)$$

Consider the case when one is willing to accept variations in the mean that are four times $SIGXB$, provided that they are sustained over relatively short periods of time. Then the tuning values using Eqs. (9-6) and (9-8) are $k = 2.0$ and $h = 2.5$.

In general, the smaller the slack variable, the more often the out-of-statistical-control signal is given. On the other hand, the smaller the value of the decision interval, the sooner an out-of-statistical-control signal is given once the mean has changed more than the acceptable value. In particular, a value of $h = k$ results in giving an out-of-statistical-control signal immediately after SH or SL becomes greater than k.

Throughout the process industries, the symmetrical distribution is a special case, because in general processes tend to be nonlinear, particularly as they are pushed toward the limits of product purity. Consider a distillation process that is designed to provide high-purity ethylene. The target purity is 98.0%, and to increase purity a certain amount above the target requires more energy per unit of product than if purity were decreased the same amount. This fact makes the gain of the process, that is, the change in purity divided by the change in energy for a given product flow, nonlinear. In this case, the distribution of the means is skewed to the left, that is, below the mean, and it has negative skewness.

Another example is moisture in instant coffee. The outlet temperature of the dryer is usually controlled by manipulating fuel flow. The moisture is sampled every hour, and when it increases above the desired target, the temperature set point is increased to remove more water and lower the moisture. The opposite is done when the moisture decreases below the desired target. As the desired target gets smaller and smaller, more fuel per kilogram of coffee is required to reduce moisture 1% below the target than to increase the same amount above the target. In this case, the moisture distribution of the means (xbars) is skewed to the right, that is, above the mean, and it has positive skewness.

The unsymmetrical CUSUM scheme recommended by the author is the appropriate one to use for both of these examples and for most, if not all, process variables. This scheme requires the determination of slack variables $k1$ and $k2$ and the decision intervals $h1$ and $h2$.

First, the variable considered for CUSUM application is analyzed using the xbar histogram described in Section 3.7. The histogram provides calculated values for *SIGXB* and the normality measures of skewness and kurtosis.

Next, choose the acceptable change in the mean as a function of the computed *SIGXB*. Then, for a positively skewed variable, as for example the moisture variable in the dryer, choose $k1$ and $h1$ using Eqs. (9-6) and (9-8). The values of $k2$ and $h2$ should be less than $k1$ and $h1$, respectively. How much less depends on how large the skewness is. In particular, the values of $k2$ and $h2$ should be chosen so that they provide the same average run length below the mean as the values of $k1$ and $h1$ provide above the mean.

Similarly, for a negatively skewed variable, such as the ethylene example above, choose $k2$ and $h2$ using Eqs. (9-6) and (9-8). The values of $k1$ and $h1$ should be less than $k2$ and $h2$, respectively. In particular, the values of $k1$ and $h1$ should be chosen so that they provide the same average run length above the mean as the values of $k2$ and $h2$ provide below the mean.

9.3 THE CUSUM AND *N* SKIP *M* FOR MINIMUM AUTOCORRELATION

The same samples used for the xbar and sigma chart of the example presented in Section 5.3 are also used to build CUSUM charts in order to show the effect of minimizing autocorrelation using the *N* skip *M* subgrouping method.

Figure 9-2 shows the CUSUM chart for the samples of Table 5-2 with the subgroup size $N = 5$ and no skipping; that is, the skip size is $M = 0$. Since the number of samples is 250, this results in a chart with 50 subgroups.

The values used for the slack variables are $k1 = k2 = 0.5$, and the values used for the decision intervals are $h1 = h2 = 4.0$. When the sum is above $h1$ or below $h2$, the variable is not in statistical control. Based on this rule, the following subgroups violated the CUSUM rule:

Rule CUSUM: 13–23, 31–50

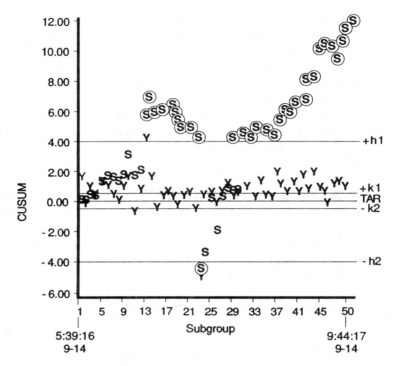

Figure 9-2   CUSUM Chart for the Samples of Table 5.2 with $N = 5$ and No Skipping.

Using the subgrouping method of $N = 5$ and skip size $M = 5$, as in Section 5.3, results in a CUSUM chart with 25 subgroups and uses all 250 samples of Table 5-2. Figure 9-3 shows the CUSUM chart for the samples of Table 5-2 with the subgroup size of $N = 5$ and skip size $M = 5$.

The same values of slack variables and decision intervals are used in this chart as those for the chart of Fig. 9-2. The following subgroups violated the CUSUM rule:

Rule CUSUM: 21–25

The results show clearly the dramatic reduction in rule violations when the skip size $M = 5$, which reduces autocorrelation substantially, is used. Moreover, the first rule violation for the chart of Fig. 9-2 occurs at subgroup 13, which represents 85 min of elapsed time from the beginning of the chart. On the other hand, the first rule violation for the chart of Fig. 9-3 occurs at subgroup 21, which represents 210 min of elapsed time from the beginning of the chart.

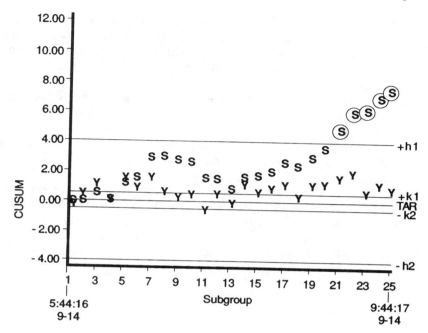

Figure 9-3   CUSUM Chart for the Samples of Table 5.2 with $N = 5$ and $M = 5$.

9.4 THE CUSUM FOR ON-DEMAND ANALYSIS AND REAL-TIME MONITORING

The CUSUM chart can be used for on-demand analysis and real-time monitoring. For on-demand analysis, the operator calls a CUSUM chart for the desired variable similar to the one shown in Fig. 9-3, with variable samples starting from the current time and going back in time long enough to obtain the configured number of subgroups or time span. This chart shows that the variable went out of statistical control at subgroup 21 and continued out of control up to subgroup 25, which contains the present time. This means the operator should have taken control action at subgroup 21 to drive the variable in the other direction toward the desired target. However, in this mode of operation the best the operator can do is to take action after the chart is displayed and indicates an out-of-statistical-control condition.

To take timely control action, real-time monitoring is required. In real-time monitoring, SH and SL are first initialized to zero and the monitor program runs periodically and checks whether a new subgroup of samples has been collected. Everytime a new subgroup is obtained using any one of the subgrouping methods of Section 3.2, a new SH and SL are computed using Eqs. (9-1) through (9-3). If SH exceeds $h1$ or SL exceeds $h2$, an alarm is given to the operator. Upon alarm acknowledgment, the SH and SL are reinitialized to zero and monitoring continues on and on, while the operator takes control action to drive the variable toward the desired target. The most timely control action is achieved when the operator takes action immediately after the monitor sets the signal indicating an out-of-statistical-control condition.

However, when the monitor signals an out-of-statistical-control condition, the operator has the option of calling on demand the CUSUM chart for the same variable or any other charts for any other variables and to move in time through the collected samples to determine what action to take. Thus it is necessary to configure CUSUM charts and have them ready to be called on demand for all variables that are under CUSUM monitoring.

So far, the control action(s) is taken by the operator upon receipt of the appropriate monitoring signal. This type of control is usually referred to as *open-loop advisory*; that is, upon monitor advice, the operator decides whether and what control action to take. On the other hand, the control action can be provided automatically on closed loop using the CUSUM Controller described in Chapter 10.

Chapter 10

CUSUM Controller*

The CUSUM Controller provides unsymmetrical, nonlinear, automatic, closed-loop feedback control action to maintain a controlled variable close to a desired target as determined by the decision parameters. It uses the cumulative deviation of the subgroup mean minus the target, divided by the sample standard deviation of the subgroup, and modifies the manipulated variable each time the accumulated variable exceeds a decision interval above or below the target. It also uses a slack variable above and below the target, which is subtracted from the cumulative deviation, and in this sense it provides slack to the calculation and the resulting control action.

The slack variable, the decision interval, and the controller gain are generally set to different values, depending on whether the cumulative sum is above or below the target. This provides for unsymmetrical nonlinear control action. Symmetrical linear control action is provided by the controller in the special case when the slack value, the decision interval, and the controller gain above the target are set equal, respectively, to their values below the target.

* United States patent pending in the name of The Foxboro Company.

10.1 CUSUM CONTROLLER CALCULATIONS

Most of the CUSUM Controller calculations are the same as those for the CUSUM chart of Section 9.1. The controller provides real-time closed-loop control action as follows:

Step 1

Ascertain that the slack variables $k1$ and $k2$ and the decision intervals $h1$ and $h2$ satisfy the conditions given by Eq. (9-4) and Eq. (9-5), respectively.
      Initialize as follows:

$$SH(j) = SH(j - 1) = SL(j) = SL(j - 1) = 0$$
$$HA = LA = 0$$

$M(j - 1)$ = actual value of the manipulated variable at time of controller mode switch from manual to automatic for bumpless transfer

Step 2

Depending on the subgrouping method chosen using Section 3.2, collect the necessary samples to form a subgroup and compute the subgroup normalized deviation from target $Y(j)$ using Eq. (9-1) and the cumulative sum high $SH(j)$ and low $SL(j)$ using Eqs. (9-2) and (9-3), respectively. Note that $Y(j)$ is positive when $XB(j)$ is above the target and negative when $XB(j)$ is below the target.
      If there is no variation, $XB(j)$ approaches a constant value and $S(j)$ approaches zero. To avoid dividing by zero, $S(J)$ is limited to the minimum sigma value $MINSIG$ as follows:

$$\text{For } S(j) < MINSIG, \qquad \text{set } S(j) = MINSIG \qquad (10\text{-}1)$$

The value of $MINSIG$ should be very small. For example, it can be chosen to be less than $SIGXB/100$.

Step 3

Check if $SH(j)$ is greater than $h1$, and if it is, set $HA = 1$. Otherwise, set $HA = 0$.
      Check if $SL(j)$ is greater than $h2$, and if it is, set $LA = 1$. Otherwise, set $LA = 0$.
      For subgroup j, either $HA = 1$ or $LA = 1$, but not both, because $Y(j)$ is added to $SH(j - 1)$ in Eq. (9-2) but subtracted from $SL(j - 1)$ in Eq. (9-3).
      In the preceding,

$$HA = \text{high alarm, with value 1 or 0}$$
$$LA = \text{low alarm, with value 1 or 0}$$

Step 4

Compute the manipulated variable for the CUSUM Controller as follows:

$$M(j) = M(j - 1) + KC1\ HA\ SH(j) - KC2\ LA\ SL(j) \qquad (10\text{-}2)$$

where

$\qquad M(j) =$ current value of the manipulated variable
$M(j - 1) =$ previous value of the manipulated variable
$\qquad KC1 =$ high gain of the controller for SH, which can have both positive and negative values
$\qquad KC2 =$ low gain of the controller for SL, which can have both positive and negative values

The values of $KC1$ and $KC2$ must both be either positive or negative.

Step 5

$\qquad$ If $HA = 1$, then set $SH(j) = 0$, $SH(j - 1) = 0$, and $HA = 0$.
$\qquad$ If $HA = 0$, then set $SH(j - 1) = SH(j)$.
$\qquad$ If $LA = 1$, then set $SL(j) = 0$, $SL(j - 1) = 0$, and $LA = 0$.
$\qquad$ If $LA = 0$, then set $SL(j - 1) = SL(j)$.
$\qquad$ Set $M(j - 1) = M(j)$.

Step 6

Repeat steps 2 through 6.

As a way of further clarification, consider the sketch shown in Fig. 10-1. The x axis is the subgroup number. The two variables plotted on the y axis are the normalized deviation from target indicated by the symbol Y and the cumulative sum high and low indicated by the symbol S. Recall that cumulative sum high $SH(j)$ is above the target TAR, and cumulative sum low $SL(j)$ is below the target TAR. Note that, although the actual value of the low sum as computed is a positive value, it is plotted as negative to indicate that the controlled variable is moving below and away from the target.

$\qquad$ Assume at subgroup 1 that the controller is placed on closed-loop control with the appropriate initialization described in step 1. As time goes on, starting with subgroup 1 up to subgroup 11, the value of Y remains above target and the cumulative sum high SH increases until it exceeds the value of the high decision interval $h1 = 2.0$ at subgroup 11, at which time the high alarm HA is set to 1, and the manipulated variable is modified in a direction to drive the controlled process variable down toward the target per step 4 and Eq. (10-2). Then the appropriate variables are initialized per step 5 and the action starts anew.

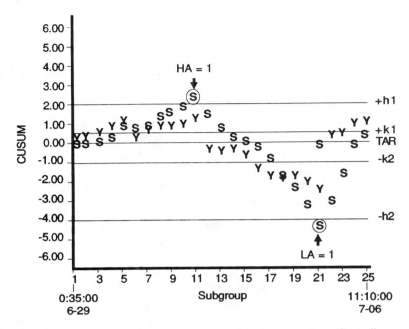

Figure 10-1 Sketch Indicating Response and Action of CUSUM Controller.

As a result of the control action, the variable Y is forced to go below the target at subgroup 12, and it remains below target until subgroup 21. At this time, the cumulative sum low SL exceeds the value of the low decision interval $h2 = 4.0$, and the low alarm LA is set to 1. The manipulated variable is then adjusted to drive the controlled process variable up toward the target.

This action is repeated as often as necessary to maintain the controlled process variable close to the target, depending on the values of the slack variables, the decision intervals, and the controller gains.

10.2 TUNING THE CUSUM CONTROLLER

The CUSUM Controller gains $KC1$ and $KC2$ need to be set so that the change in the manipulated variable is sufficient to drive the controlled variable to its target while maintaining control loop stability. In particular, the gains should be set large enough to overshoot the target from above and below in order to avoid offset. Both gains have the same sign. When the sign is positive, the controller provides increase/increase action in that the manipulated variable is increased when the actual variable is above the target ($Y > 0$). When the sign is negative, the controller provides increase/decrease action in that the manipulated variable is decreased when the actual variable is above the target ($Y > 0$).

The slack value $k1$, the decision interval $h1$, and the controller gain $KC1$ determine the speed of response for the controlled variable above the target ($Y > 0$). The slack value $k2$, the decision interval $h2$, and the controller gain $KC2$ determine the speed of response for the controlled variable below the target ($Y < 0$). The smaller the value of $k1$, $h1$, $k2$, and $h2$, the more frequently the manipulated variable is changed, which results in tighter control. Of course, their values can be made so small as to force the controller to take action as often as a new subgroup of controlled variable samples is obtained.

The tuning recommendations for $k1$, $h1$, $k2$, and $h2$ given in Section 9.2 for the CUSUM chart apply to the CUSUM Controller as well. However, since the control is closed loop automatic, smaller values can be chosen to provide control action more often and keep the variable closer to the target, as compared to when the action is taken by the operator.

10.3 CUSUM CONTROLLER APPLICATIONS

Figure 10-2 shows a simplified dryer control system. In this dryer, fuel is burned in the combustion chamber to heat ambient air. The hot air enters the dryer and comes in contact with the wet feed. As the hot air and feed move parallel to each other, heat from the hot air is transferred to the wet material and moisture is removed into the air. Moist air is exhausted to the atmosphere, and the dried product is removed at the discharge end of the dryer.

The control system consists of three loops cascaded to each other. The innermost loop and the fastest is the fuel flow controller FC, which uses the fuel flow measurement provided by transmitter FT and controls fuel flow. The next loop is the exhaust air temperature controller TC, which is slower than the flow loop and uses the exhaust air temperature measurement provided by transmitter TT and manipulates fuel flow to maintain the temperature at its set point.

The set point of the temperature controller in turn is manipulated by the CUSUM Moisture Controller, which uses the moisture measurement provided by the moisture transmitter MT to control moisture at the desired target. This loop is the slowest. For comparison purposes, the flow controller responds in a couple of seconds, the temperature controller in a couple of minutes, and the moisture controller in hours.

The CUSUM Moisture Controller can also be used to provide the appropriate control action when the moisture measurement comes from the laboratory and is entered by the operator, as opposed to being automatically sampled by the moisture transmitter.

In this example, the CUSUM Controller is at the top of the cascade and controls moisture, which is the quality variable. This is a typical use of the CUSUM Controller as the top-level controller in the cascade. It usually controls quality variables or variables that infer quality. The quality variables can be au-

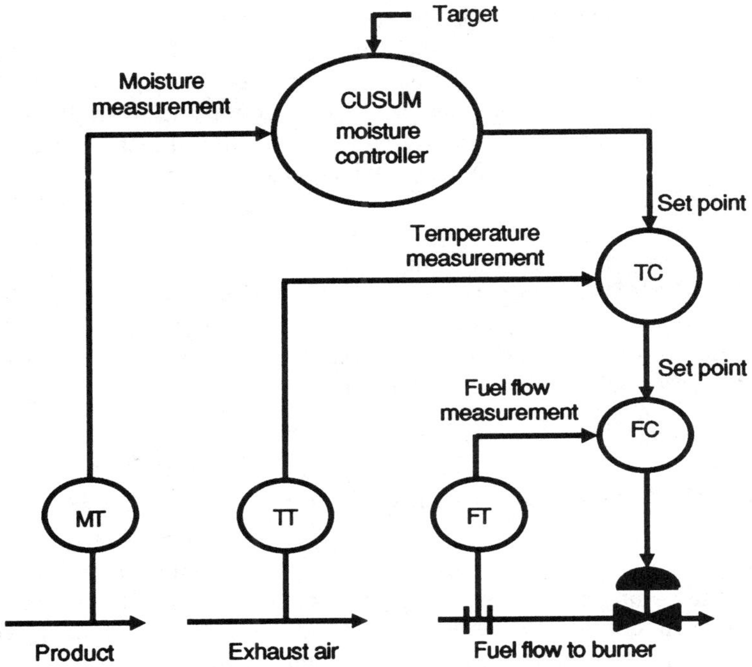

Figure 10-2   CUSUM Controller Application Diagram.

tomatically measured by on-line sensors, or their values can be entered manually by the operator based on product samples analyzed in the laboratory.

Traditional controllers for quality variables control them to a target or set point that is a single value, while the CUSUM Controller controls them to a target with slack around it. The traditional controller takes some control action every sample and tries to drive the process variable to the target every sample period. This may cause more variation than the CUSUM Controller for nonlinear processes with long residence times, particularly at the highest level of cascaded control.

Chapter 11

Optimum Setpoint Controller*

The purpose of the Optimum Setpoint Controller is to maintain the set point (target) of a controlled variable as close as possible to a specification limit based on its variation. This is possible when the specification is one sided; for example, the 3-sigma moisture variation of instant coffee should not exceed a high moisture limit of 3.0%. Thus, measuring the sigma of the moisture samples, one can maintain the moisture set point at a value that is always 3-sigma below the limit.

An operator set moisture set point is usually set to a lower value to guarantee product on specification under worst-case conditions, that is, the largest variation expected. The Optimum Setpoint Controller, on the other hand, shifts the set point automatically, which results in optimizing the process by avoiding over-drying and selling the maximum amount of water allowed, whenever the variation is less than the largest expected.

There are situations when a low specification limit is required, in which case the variable approaches it from above; for example, the 3-sigma variation of ethylene composition in a distillation column should not go below 98%.

To guarantee that a specific percentage of samples falls above the high specification limit or below the low specification limit, the distribution of the samples must be nearly normal. If the individual samples do not form an acceptably normal

* United States patent pending in the name of The Foxboro Company.

distribution, then the subgroup means can be used instead of individual samples to compute the optimum set point. This is done, because the mean values tend toward the normal distribution as described in Section 3.2. Moreover, subgrouping method N skip M can be used to minimize correlation as described in Section 4.3.

Thus the optimum set point calculation is based on a specification limit and *SIGXB*, xbar sigma, which is the standard deviation of the subgroup mean values. The calculated normality measures for the subgroup mean values discussed in Section 3.7 can be used optionally to either bypass the optimum set point calculation when the normality criterion is not satisfied or to perform the computation and set an alarm for the operator.

11.1 OPTIMUM SETPOINT CONTROLLER CALCULATIONS

The controller computes the necessary closed-loop control action as follows: Initially, it obtains the required number of individual samples for a given variable from the real-time database starting with the current time. This number depends on the number of samples per subgroup, the number of subgroups, and the subgrouping method used, as described in Section 3.2.

The optimum set point is computed by

$$OSPT = SL + KPAR\ SIGXB \qquad (11\text{-}1)$$

where

$OSPT$ = optimum set point
SL = specification limit
$KPAR$ = parameter with positive or negative value
$SIGXB$ = xbar sigma

For subgroup size less than or equal to 10, the value of *SIGXB* is computed using Eqs. (3-2), and (3-8). For subgroup size greater than 10, the value of *SIGXB* is computed using Eqs. (3-10) and (3-11).

The values of $R(j)$, Eq. (3-9), or $S(j)$, Eq. (3-3), for $j = 1, \ldots, NS$ are saved in memory. The data collection continuous, and when enough samples have been collected to form a new subgroup, a new value for R or S is computed using Eq. (3-2) or (3-3). The new value then replaces the oldest value for R or S in the memory-stored array of NS values, and *SIGXB* is computed anew. The new value of *SIGXB* is then used to compute the new optimum set point using Eq. (11-1). This process is repeated everytime a new subgroup of values is collected.

Optionally, the standard normality measures of skewness and kurtosis can be computed everytime a new subgroup is obtained. If either or both normality measures are within acceptable limits, then the optimum set point calculation is either skipped or it is made and an alarm is given to the operator.

The value of *KPAR* is positive when approaching the specification limit from

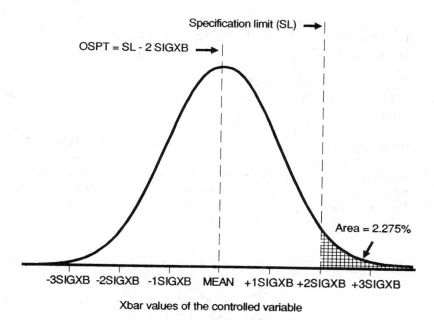

Figure 11-1   The Normal Curve and Specification Limit Approached from Below.

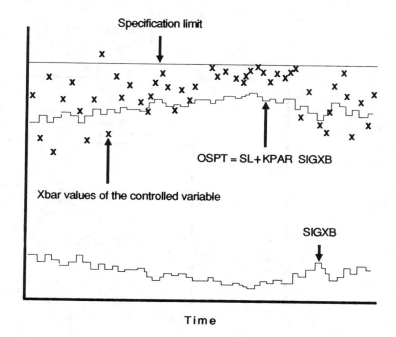

Figure 11-2   Sketch for Optimum Set Point and Xbar and SIGXB Trends.

above, and it is negative when approaching the specification limit from below. The parameter *KPAR* sets the percent of product failing to meet specification. The normal curve of Fig. 11-1 shows a specification limit approached from below with *KPAR* = −2. For this value of *KPAR*, more than 2.275% of product is above the specification limit. For *KPAR* = −1, the percent above the specification limit is 15.865%, and for *KPAR* = −3, the percent above the specification limit is 0.135%.

Figure 11-2 provides a sketch of time trends for the variables involved in the computation of the optimum set point. Initially, the variation is large, as indicated by the xbar and *SIGXB* trends of the controlled variable. As *SIGXB* decreases, the optimum set point moves closer to the specification limit as defined by *KPAR*. As *SIGXB* increases, the optimum set point moves farther away from the specification limit. The specification limit in this case is approached from below.

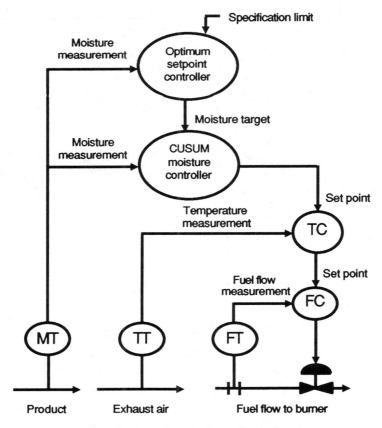

Figure 11-3 Optimum Setpoint Controller Application.

11.2 OPTIMUM SETPOINT CONTROLLER APPLICATIONS

The target value in the CUSUM Moisture Controller of the dryer control system shown in Fig. 10-2 can be computed using the Optimum Setpoint Controller as shown in Fig. 11-3. Of course, the CUSUM Controller can be any type of controller. Everytime a new subgroup of moisture measurements is obtained, the Optimum Setpoint Controller computes a new value for the moisture target.

As discussed in the beginning of this chapter, the moisture specification limit for instant coffee is 3%. Figure 11-2 can be used to explain the Optimum Setpoint Controller action for the moisture target. Let the xbar values of the controlled variable be the moisture xbar values, and let the computed $SIGXB$ vary as shown. Then the optimum moisture target is computed as shown by the curve of the optimum set point.

The Optimum Setpoint Controller can be combined with the CUSUM Moisture Controller or any other type of moisture controller, such as, for example, the traditional proportional + integral + derivative (PID) to form a single entity. The combined controller then uses the controlled variable measurement to compute the optimum set point and adjusts the manipulated variable to maintain the controlled variable at the optimum set point.

Chapter 12

P and NP Charts for Fraction and Number Defective

In all processes, many quality characteristics manifest themselves as attributes. For instance, a sample of inspected items can be classified as items accepted or conforming to specifications and items rejected or not conforming to specifications. For example, out of a sample of 100 barrels of polymer, 95 were accepted by the quality assurance and shipped to the customer and 5 were rejected. The fraction defective is $\frac{5}{100} = 0.05$ or 5%, while the number defective is 5.

The P chart is a plot of fraction defective or percent defective versus subgroup number or time, together with the mean value and the upper and lower control limits. The P chart is used when the number of tested items or subgroup size varies from sample to sample. The variable subgroup size causes the control limits to vary from subgroup to subgroup.

The fraction defective P follows the binomial or Bernoulli distribution (see reference 4). Its mean value and the subgroup size determine the subgroup sigma.

The NP chart is a plot of number defective versus subgroup number or time, together with the mean value and the upper and lower control limits. The NP chart is used when the number of tested items or subgroup size is constant from sample to sample, which makes the control limits the same for all subgroups.

12.1 CALCULATIONS AND EXAMPLE *P* CHART

The fraction defective per subgroup is

$$P(j) = \frac{ND(j)}{SZ(j)}, \qquad j = 1, \ldots, NS \qquad (12\text{-}1)$$

The fraction defective mean is

$$PB = \frac{\sum\limits_{j=1}^{NS} ND(j)}{\sum\limits_{j=1}^{NS} SZ(j)} \qquad (12\text{-}2)$$

The fraction defective follows the binomial distribution, for which the sigma per subgroup is

$$S(j) = \sqrt{\frac{PB(1 - PB)}{SZ(j)}}, \qquad j = 1, \ldots, NS \qquad (12\text{-}3)$$

The central line and control limits are computed as follows:

$$CL = PB \qquad (12\text{-}4)$$

$$UCL(j) = PB + K1\,S(j), \qquad j = 1, \ldots, NS \qquad (12\text{-}5)$$

$$LCL(j) = PB - K1\,S(j), \qquad j = 1, \ldots, NS \qquad (12\text{-}6)$$

$$\text{If } LCL(j) < 0, \text{ then set } LCL(j) = 0, \qquad j = 1, \ldots, NS \qquad (12\text{-}7)$$

In Eqs. (12-1) through (12-7),

$P(j)$ = *j*th subgroup fraction defective
$ND(j)$ = *j*th subgroup number defective
$SZ(j)$ = *j*th subgroup size
PB = fraction defective mean
$S(j)$ = *j*th subgroup sigma
CL = central line
$UCL(j)$ = *j*th subgroup upper control limit
$LCL(j)$ = *j*th subgroup lower control limit

When it is desired to plot the fraction defective as percent, the values of $P(j)$, PB, $UCL(j)$, and $LCL(j)$ are multiplied by 100.

Let *OPB* be the official fraction defective mean. Then the official sigma values for all subgroups are computed using *OPB* in Eq. (12-3). These official values of sigma, the official mean *OPB*, the calculated mean *PB*, and the calculated values of sigma using *PB* in Eq. (12-3) are used to obtain any one of the four combinations for rule evaluations and limits as described in Section 2.7.

TABLE 12-1   *P* CHART SAMPLES AND CALCULATIONS

Subgroup	Date	Time	SZ	ND	P	S	UCL	LCL
1	04/01	08:10:00	200	34	0.1700	0.0214	0.1660	0.0377
2	04/02	06:07:00	350	20	0.0571	0.0162	0.1504	0.0534
3	04/03	07:12:00	240	38	0.1583	0.0195	0.1604	0.0433
4	04/04	08:10:00	440	27	0.0614	0.0144	0.1451	0.0586
5	04/05	09:05:00	290	32	0.1103	0.0178	0.1551	0.0586
6	04/06	10:18:00	270	30	0.1111	0.0184	0.1571	0.0466
7	04/07	11:20:00	180	14	0.0777	0.0225	0.1695	0.0342
8	04/08	12:14:00	220	37	0.1681	0.0204	0.1630	0.0407
9	04/09	13:15:00	200	43	0.2150	0.0214	0.1660	0.0377
10	04/10	14:22:00	330	28	0.0848	0.0166	0.1518	0.0519
11	04/11	15:25:00	210	29	0.1381	0.0209	0.1645	0.0392
12	04/12	16:18:00	310	32	0.1032	0.0172	0.1534	0.0503
13	04/13	17:15:00	390	35	0.0897	0.0153	0.1478	0.0559
14	04/14	18:12:00	250	30	0.1200	0.0191	0.1592	0.0445
15	04/15	19:14:00	170	37	0.2176	0.0232	0.1714	0.0323
16	04/16	20:20:00	440	26	0.0591	0.0144	0.1451	0.0586
17	04/17	07:12:00	540	51	0.0944	0.0130	0.1409	0.0628
18	04/18	08:15:00	130	13	0.1000	0.0265	0.1814	0.0223
19	04/19	09:05:00	360	45	0.1250	0.0159	0.1497	0.0540
20	04/20	10:17:00	480	55	0.1146	0.0138	0.1433	0.0604
21	04/21	11:20:00	190	34	0.1789	0.0219	0.1677	0.0360
22	04/22	12:04:00	410	19	0.0463	0.0149	0.1467	0.0570
23	04/23	13:15:00	250	11	0.0440	0.0191	0.1592	0.0445
24	04/24	14:12:00	310	16	0.0516	0.0172	0.1534	0.0503
25	04/25	15:20:00	380	32	0.0842	0.0155	0.1484	0.0553

Table 12-1 provides sample values for number defective *ND*, subgroup size *SZ*, and the resulting calculations for *P, S, UCL*, and *LCL*. Other parameters and calculations are as follows:

$$NS = 25 \quad K1 = 3.0 \quad PB = 0.1019 \quad CL = 0.1019$$

Figure 12-1 shows the *P* chart for this example.

12.2 CALCULATIONS AND EXAMPLE *NP* CHART

For the *NP* chart the subgroup size is constant for all subgroups. The number defective mean is

$$NPB = \frac{\sum_{j=1}^{NS} ND(j)}{NS} \tag{12-8}$$

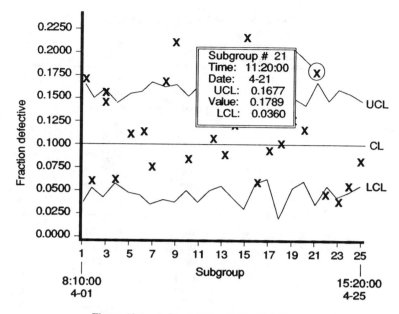

Figure 12-1   *P* Chart Using Table 12.1 Data.

The number of defective items also follows the binomial distribution, for which the sigma is

$$\text{sigma} = \sqrt{NPB\left(1 - \frac{NPB}{N}\right)} \tag{12-9}$$

The central line and control limits are computed as follows:

$$CL = NPB \tag{12-10}$$

$$UCL = NPB + K1\ \text{sigma} \tag{12-11}$$

$$LCL = NPB - K1\ \text{sigma} \tag{12-12}$$

$$\text{If } LCL < 0, \text{ then set } LCL = 0 \tag{12-13}$$

where

NPB = number defective mean

Let $ONPB$ be the official number defective mean. Then the official sigma is computed using $ONPB$ in Eq. (12-9). The official value of sigma, the official mean $ONPB$, the calculated mean NPB, and the calculated sigma using NPB in Eq. (12-9) are used to obtain any one of the four combinations for rule evaluations and limits as described in Section 2.7.

TABLE 12-2 *NP* CHART SAMPLES, SUBGROUP
SIZE *N* = 100

Subgroup	Date	Time	*ND*
1	04/01	08:10:00	14
2	04/02	06:07:00	7
3	04/03	07:12:00	16
4	04/04	08:10:00	11
5	04/05	09:05:00	13
6	04/06	10:18:00	12
7	04/07	11:20:00	4
8	04/08	12:14:00	15
9	04/09	13:15:00	18
10	04/10	14:22:00	11
11	04/11	15:25:00	12
12	04/12	16:18:00	13
13	04/13	17:15:00	14
14	04/14	18:12:00	12
15	04/15	19:14:00	15
16	04/16	20:20:00	10
17	04/17	07:12:00	22
18	04/18	08:15:00	4
19	04/19	09:05:00	19
20	04/20	10:17:00	23
21	04/21	11:20:00	14
22	04/22	12:04:00	7
23	04/23	13:15:00	2
24	04/24	14:12:00	5
25	04/25	15:20:00	13

Table 12-2 provides sample values for number defective *ND*. Other parameters and calculations are as follows:

$$N = 100 \quad\quad UCL = 22.07 \quad\quad \text{sigma} = 3.277 \quad\quad K1 = 3.0$$
$$NPB = 12.24 \quad\quad NS = 25 \quad\quad LCL = 2.408 \quad\quad CL = 12.24$$

Figure 12-2 shows the *NP* chart for this example.

12.3 WHICH RULES APPLY AND WHICH ONES TO USE

In general, only rule 1 of Table 2-3 is used to evaluate the in-statistical-control state of *P* and *NP* charts. However, for the *P* chart the binomial can be approximated by the normal distribution. According to reference 4, a good rule of thumb

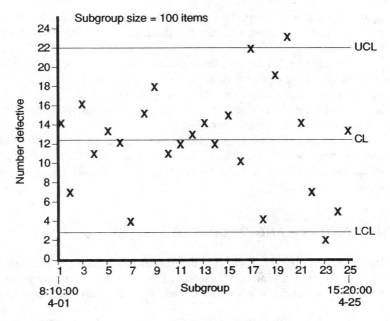

Figure 12-2 *NP* Chart Using Table 12.2 Data.

is to use the normal curve as an approximation to the binomial when the following condition is satisfied:

$$P(j)\ SZ(j) \geq 5, \qquad j = 1, \ldots, NS \qquad (12\text{-}14)$$

When the above relationship is satisfied for all subgroups, then all 11 rules of Table 2-3 can be used to evaluate the in-statistical-control state of the fraction defective attribute variable. This is true of Fig. 12-1, where all rules apply and rule violations are shown by the larger symbols.

For the *NP* chart Eq. (12-14) becomes

$$NPB \geq 5 \qquad (12\text{-}15)$$

When this relationship is satisfied, all 11 rules of Table 2-3 can be used to evaluate the in-statistical-control state of the number defective attribute variable.

Further examination of Eqs. (12-14) and (12-15) reveals that, as the fraction or number defective gets smaller and smaller, the subgroup size must be increased in order to use the normal approximation and apply any or all rules of Table 2-3.

Consider Fig. 12-1 and ask what it means when the process is not in statistical control and the sample is below the lower control limit. In particular, what does it mean when the sample shows zero defective? For manual samples, it could mean that the analysis was not done correctly by the operator or, for automatic sensors, that the sensor had failed. Or if improvements have been made to the process, perhaps some samples for the chart are from the old and some from the new process.

Although all rules can be applied when the normal approximation condition is satisfied, some of them make more sense to use than others. For example, when the mean remains the same for a long time, rule 5, gradual upward change in level, and rule 7, upward trend, can be particularly useful for monitoring the deterioration of process operation.

Typically, fraction and number defective attribute variables are not under closed-loop control. Therefore, as in the case of the xbar and range chart described in Section 5.6 for real-time monitoring, rules 1 and 3 can be used with official mean and sigma.

12.4 WHEN TO USE *P* AND *NP* CHARTS AND WHY

In general, the *P* or *NP* chart is used to monitor final products in process industry or manufacturing plants. Of course, final products for one plant can be raw materials for another, in which case the *P* or *NP* chart can be used to monitor them as well.

Consider a chemical plant that produces polymer. The polymer is monitored using a *P* chart for the fraction defective of tank cars or barrels produced in a given period of time. The polymer is the raw material for plastic products used directly by the consumer or as parts of other products, such as automobiles and refrigerators. Let one of these products be a plastic part for an automobile door; then a *P* chart can be built using the fraction defective of the plastic parts for the automobile door. Finally, one can build a *P* chart for the fraction of defective automobiles of a particular model and for a given period of time. Another example is a pulp and paper mill for which a *P* chart can be built based on the fraction defective of rolls of paper produced in a given period of time, say, each day. For these examples, the *P* chart is replaced by the *NP* chart when the subgroup size is constant.

If the *P* or *NP* chart indicates that the fraction or number defective is not in statistical control because it violates one or more of the chosen rules, additional analysis is necessary to identify and solve the problem. Since the *P* and *NP* charts typically show final product results, other charts for variables that determine product conditions during processing must be used. Among them are the xbar and range or sigma, individuals, and CUSUM charts, which are very effective in monitoring and analyzing process conditions. Examples of variables that show conditions during processing are feed flows and compositions, ratio of feed to catalyst, catalyst activity, temperature, and residence time.

Chapter 13

C and *U* Charts
for Defects
and Defects per Unit

Another type of attribute is the number of defects in a sample, for example, the number of holes in a yard or square meter of cloth or the number of blemishes on a roll of paper. The *C* chart applies in this case. It is a plot of number of defects versus subgroup number or time, together with the mean value and the upper and lower control limits. The *C* chart is used when the unit size is constant, for example, the number of defects in a yard of cloth, where the unit size is a yard of cloth every time. The number of defects *C* follows the Poisson distribution and its mean value determines the sigma (see reference 4).

The *U* chart is a plot of defects per unit and is used when the unit size varies from subgroup to subgroup. An example is the number of defects per yard of cloth where the unit size is 1.0 yards of cloth for the first subgroup, 1.35 yards of cloth for the second subgroup, and so on. The variable unit size causes the control limits to vary from subgroup to subgroup. The subgroup unit size and the defects per unit mean determine the subgroup sigma.

13.1 CALCULATIONS AND EXAMPLE *C* CHART

For the *C* chart, the unit size is constant for all subgroups. The number of defects mean is

$$CB = \frac{\sum\limits_{j=1}^{NS} C(j)}{NS} \qquad (13\text{-}1)$$

The number of defects follows the Poisson distribution, for which the sigma is

$$\text{sigma} = \sqrt{CB} \qquad (13\text{-}2)$$

The central line and control limits are computed as follows:

$$CL = CB \qquad (13\text{-}3)$$

$$UCL = CB + K1 \text{ sigma} \qquad (13\text{-}4)$$

$$LCL = CB - K1 \text{ sigma} \qquad (13\text{-}5)$$

$$\text{If } LCL < 0, \text{ then set } LCL = 0 \qquad (13\text{-}6)$$

In Eqs. (13-1) through (13-6),

$$C(j) = j\text{th subgroup number of defects}$$

$$CB = \text{number of defects mean}$$

Let *OCB* be the official number of defects mean. Then the official sigma is computed using *OCB* in Eq. (13-2). The official value of sigma, the official mean *OPB*, the calculated mean *PB*, and the calculated sigma using *CB* in Eq. (13-2) are used to obtain any one of the four combinations for rule evaluations and limits as described in Section 2.7.

Table 13-1 provides sample values for number of defects *C*. Other parameters and calculations are as follows:

size = 1.5	NS = 25	K1 = 3.0
CB = 13.84	sigma = 3.720	CL = 13.84
UCL = 25.00	LCL = 2.679	

Figure 13-1 shows the *C* chart for this example.

13.2 CALCULATIONS AND EXAMPLE *U* CHART

The number of defects per unit per subgroup is

$$U(j) = \frac{C(j)}{SZ(j)} \qquad j = 1, \ldots, NS \qquad (13\text{-}7)$$

TABLE 13-1   *C* CHART SAMPLES, Size = 1.5

Subgroup	Date	Time	*C*
1	04/01	08:10:00	15
2	04/02	06:07:00	8
3	04/03	07:12:00	17
4	04/04	08:10:00	13
5	04/05	09:05:00	15
6	04/06	10:18:00	14
7	04/07	11:20:00	6
8	04/08	12:14:00	15
9	04/09	13:15:00	19
10	04/10	14:22:00	14
11	04/11	15:25:00	13
12	04/12	16:18:00	15
13	04/13	17:15:00	16
14	04/14	18:12:00	14
15	04/15	19:14:00	17
16	04/16	20:20:00	12
17	04/17	07:12:00	23
18	04/18	08:15:00	5
19	04/19	09:05:00	21
20	04/20	10:17:00	25
21	04/21	11:20:00	16
22	04/22	12:04:00	9
23	04/23	13:15:00	3
24	04/24	14:12:00	7
25	04/25	15:20:00	14

The number of defects per unit mean is

$$UB = \frac{\sum_{j=1}^{NS} C(j)}{\sum_{j=1}^{NS} SZ(j)} \qquad (13\text{-}8)$$

The number of defects per unit also follow the Poisson distribution, for which the sigma per subgroup is

$$S(j) = \sqrt{\frac{UB}{SZ(j)}}, \qquad j = 1, \ldots, NS \qquad (13\text{-}9)$$

The central line and control limits are computed as follows:

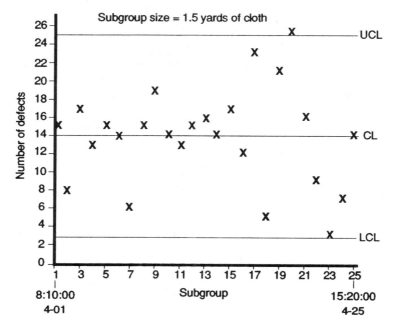

Figure 13-1   *C* Chart Using Table 13.1 Data.

$$CL = UB \tag{13-10}$$

$$UCL(j) = UB + K1\, S(j), \qquad j = 1, \ldots, NS \tag{13-11}$$

$$LCL(j) = UB - K1\, S(j), \qquad j = 1, \ldots, NS \tag{13-12}$$

$$\text{If } LCL(j) < 0, \text{ then set } LCL(j) = 0, \qquad j = 1, \ldots, NS \tag{13-13}$$

In Eqs. (13-7) through (13-13),

$U(j) = j$th subgroup defects per unit
$SZ(j) = j$th subgroup size
$UB = $ defects per unit mean
$S(j) = j$th subgroup sigma
$CL = $ central line
$UCL(j) = j$th subgroup upper control limit
$LCL(j) = j$th subgroup lower control limit

Let *OUB* be the official defects per unit mean. Then the official sigma values for all subgroups are computed using *OUB* in Eq. (13-9). These official values of sigma, the official mean *OUB*, the calculated mean *UB*, and the calculated values of sigma using *UB* in Eq. (13-9) are used to obtain any one of the four combinations for rule evaluations and limits as described in Section 2.7.

Table 13-2 provides sample values for number of defects *C*, subgroup size

TABLE 13-2 *U* CHART SAMPLES AND CALCULATIONS

Subgroup	Date	Time	*SZ*	*C*	*U*	*S*	*UCL*	*LCL*
1	04/01	08:10:00	1.5	32	21.333	2.8325	20.532	3.5370
2	04/02	06:07:00	2.7	19	7.037	2.1112	18.368	5.7008
3	04/03	07:12:00	1.8	36	20.000	2.5857	19.792	4.2774
4	04/04	08:10:00	3.4	25	7.353	1.8814	17.679	6.3904
5	04/05	09:05:00	2.2	30	13.636	2.3389	19.051	5.0179
6	04/06	10:18:00	2.1	27	12.857	2.3939	19.216	4.8528
7	04/07	11:20:00	1.4	11	7.857	2.9319	20.830	3.2388
8	04/08	12:14:00	1.7	34	20.000	2.6607	20.016	4.0525
9	04/09	13:15:00	1.5	40	26.667	2.8325	20.532	3.5370
10	04/10	14:22:00	2.5	25	10.000	2.1940	18.617	5.4524
11	04/11	15:25:00	1.6	27	16.875	2.7425	20.262	3.8068
12	04/12	16:18:00	2.4	29	12.083	2.2393	18.752	5.3166
13	04/13	17:15:00	3.1	33	10.645	1.9703	17.945	6.1236
14	04/14	18:12:00	1.9	28	14.737	2.5167	19.585	4.4843
15	04/15	19:14:00	1.3	34	26.154	3.0426	21.162	2.9067
16	04/16	20:20:00	3.4	24	7.059	1.8814	17.679	6.3904
17	04/17	07:12:00	4.1	45	10.975	1.7133	17.174	6.8947
18	04/18	08:15:00	1.1	11	10.000	3.3076	21.957	2.1116
19	04/19	09:05:00	2.7	41	15.185	2.1112	18.368	5.7008
20	04/20	10:17:00	3.7	49	13.243	1.8035	17.445	6.6240
21	04/21	11:20:00	1.5	31	20.667	2.8325	20.532	3.5370
22	04/22	12:04:00	3.2	17	5.313	1.9393	17.853	6.2167
23	04/23	13:15:00	1.9	9	4.737	2.5167	19.585	4.4843
24	04/24	14:12:00	2.4	13	5.417	2.2393	18.752	5.3166
25	04/25	15:20:00	2.9	28	9.655	2.0371	18.146	5.9235

SZ, and the resulting calculations for *U*, *S*, *UCL*, and *LCL*. Other parameters and calculations are as follows:

$$NS = 25 \quad K1 = 3.0 \quad UB = 12.0345 \quad CL = 12.0345$$

Figure 13-2 shows the *U* chart for this example.

13.3 WHICH RULES APPLY AND WHICH ONES TO USE

In general, only rule 1 of Table 2-3 is used to evaluate the in-statistical-control state of *C* and *U* charts. However, for the *C* chart the Poisson can be approximated by the normal distribution. According to reference 4, a good rule of thumb is to

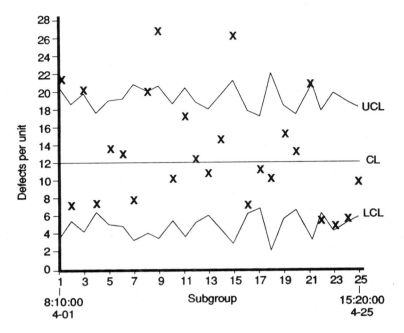

Figure 13-2   *U* Chart Using Table 13.2 Data.

use the normal curve as an approximation to the Poisson when the following condition is satisfied:

$$CB \geq 10 \qquad (13\text{-}14)$$

When the above relationship is satisfied, all 11 rules of Table 2–3 can be used to evaluate the in-statistical-control state of the fraction defective attribute variable.

For the U chart, Eq. (13–14) becomes

$$U(j)\,SZ(j) \geq 10, \qquad j = 1, \ldots, NS \qquad (13\text{-}15)$$

When this condition is satisfied for all subgroups, all 11 rules of Table 2-3 can be used to evaluate the in-statistical-control state of the fraction defective attribute variable.

Further examination of Eqs. (13–14) and (13-15) reveals that, as the number of defects or defects per unit gets smaller and smaller, the subgroup size must be increased in order to use the normal approximation and apply any or all rules of Table 2-3.

Although all rules can be applied when the normal approximation condition is satisfied, some of them make more sense to use than others. For example, when the mean remains the same for a long time, rule 5, gradual upward change in level, and rule 7, upward trend, can be particularly useful for monitoring the deterioration of process operation.

Typically, number of defects and defects per unit attribute variables are not under closed-loop control. Therefore, as in the case of the xbar and range chart described in Section 5.6 for real-time monitoring, rules 1 and 3 can be used with official mean and sigma.

13.4 WHEN TO USE *C* AND *U* CHARTS AND WHY

In general, the *C* or *U* charts, much like the *P* and *NP* charts, are used to monitor final products in process industry or manufacturing plants. Of course, final products of one plant can be raw materials for another, in which case the *C* or *U* charts can be used to monitor them as well.

Consider a textile plant that makes fabric. The fabric is monitored using a *U* chart for defects per yard of fabric made in a given period of time. The fabric is the raw material for a clothes manufacturer that makes garments that are used directly by the consumer. Let one type of garment be blouses. Then a *P* chart can be built using the fraction defective of the blouses.

If the *C* or *U* chart indicates that the number of defects or defects per unit is not in statistical control because it violates one or more of the chosen rules, additional analysis is necessary to identify and solve the problem. Since the *C* and *U* charts typically show final product results, other charts for variables that determine product conditions during processing must be used. Among them are the xbar and range or sigma, individuals, and CUSUM charts, which are very effective in monitoring and analyzing process conditions. Examples of variables that show conditions during processing are dye flow and color, ratio of dye to fiber, pressure, and temperature.

Chapter 14

The Pareto Diagram and the 80/20 Rule

The Pareto diagram described in reference 9 is the first step in determining what to work on in order to make process improvements. It can also be used to determine what effect a set of improvements has had on a given process. The basis for it is the observation that 80% of defects are attributed to 20% of the causes, which is known as Pareto's law or the 80/20 rule.

The Pareto diagram plots the number of occurrences of product or service rejection versus cause, prioritized according to size. That is, the cause with the largest number of occurrences is first, the one with the second largest number of occurrences is second, and so on.

Figure 14-1 shows a typical Pareto diagram. The bar graph is associated with the left Y axis, which indicates the number of occurrences. The point graph is associated with the right Y axis, which indicates percent contribution. The X axis shows the causes. The height of each bar is proportional to the number of occurrences for a given cause in relation to the total number of occurrences due to all causes.

The utility of the Pareto diagram is enhanced by providing weighting coefficients for each cause. The coefficients are used to convert the number of occurrences to any desirable unit of measure, for example, cost in dollars or any other currency.

For purposes of clarity, the discussion so far has been limited to occurrences

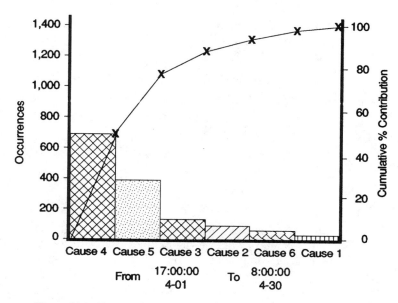

Figure 14-1   Pareto Diagram for Occurrences Using Table 14.1 Data.

of rejection and their associated causes. However, more generality is achieved by talking about variables, instead of occurrences of rejection and their associated causes. Combining this with the weighting coefficients for each variable, one can represent any desired quantity on the left Y axis of the Pareto diagram.

The real-time Pareto diagram presented here can not only plot a diagram for single values of each occurrence or variable, but also the sum of occurrences or variable values for a given period of time, say 30 days. Thus one can build a Pareto diagram for a day, a week, a month, or any other time span desired for analysis.

14.1 CALCULATIONS AND EXAMPLE PARETO DIAGRAMS

The real-time database collects the variable values for each cause. The real-time Pareto diagram can be configured to access either a single variable value for each cause or a number of values within a given time span, starting with the current time. If more than one value for each cause is retrieved, the sum of the values is computed to obtain the total for each variable and its associated cause.

Let $X(1), \ldots, X(NCR)$ represent a single value or the sum of values for a desired time period for variables associated with causes 1 to NCR. Now let $W(1), \ldots, W(NCR)$ be the corresponding weighting coefficients for causes 1 to NCR. Then the weighted variables are computed as follows:

$$WX(i) = W(i) \, X(i), \qquad i = 1, \ldots, NCR \qquad (14\text{-}1)$$

Now let $V(1), \ldots, V(NCR)$ represent the weighted variables $WX(1), \ldots, WX(NCR)$ prioritized according to size, with $V(1)$ the largest and $V(NCR)$ the smallest. That is,

$$V(1) < V(2), \ldots, < V(NCR) \tag{14-2}$$

The percent contribution for each cause is computed as follows:

$$SWV = \sum_{i=1}^{NCR} V(i) \tag{14-3}$$

$$PC(i) = \frac{100V(i)}{SWV}, \qquad i = 1, \ldots, NCR \tag{14-4}$$

In Eqs. (14-1) through (14-4),

NCR = number of causes for rejection
$X(i)$ = single value or sum of values for variable associated with cause i
$W(i)$ = weighting coefficient for cause i
$WX(i)$ = weighted variable $X(i)$
$V(i)$ = weighted variable with priority i
SWV = sum of prioritized weighted variables
$PC(i)$ = percent contribution of cause with priority i

Table 14-1 provides 30 sample values for six causes for rejection, one for each day of the month. The sample values represent the number of occurrences for each cause. Variable $X(1)$ is associated with cause 1, $X(2)$ with cause 2, ..., $X(6)$ with cause 6. Variables $V(1)$ to $V(6)$ represent the sum of the 30 samples for each $X(1)$ to $X(6)$ prioritized according to size, with $V(1)$ the largest. The weighting coefficients were set to their default values 1.0. Variables $PC(1)$ to $PC(6)$ represent the percent contributions of $V(1)$ to $V(6)$, respectively. Other parameters used are as follows:

$$NCR = 6$$

$$W(1) = W(2) = W(3) = W(4) = W(5) = W(6) = 1.0$$

The Pareto diagram for the data of Table 14-1 is shown in Fig. 14-1.

Now consider the same data and the desire to convert them to cost in dollars and then build the Pareto diagram. Let the weighting coefficients be

$$W(1) = \$80/\text{occurrence, associated with cause 1}$$

$$W(2) = \$85/\text{occurrence, associated with cause 2}$$

$$W(3) = \$90/\text{occurrence, associated with cause 3}$$

$$W(4) = \$3/\text{occurrence, associated with cause 4}$$

$$W(5) = \$4/\text{occurrence, associated with cause 5}$$

$$W(6) = \$30/\text{occurrence, associated with cause 6}$$

TABLE 14-1   PARETO DIAGRAM SAMPLES AND CALCULATIONS

Subgroup	Date	Time	$X(1)$	$X(2)$	$X(3)$	$X(4)$	$X(5)$	$X(6)$
1	04/01	08:10:00	2	2	5	44	10	2
2	04/02	06:07:00	1	3	3	20	12	2
3	04/03	07:12:00	0	3	4	19	13	1
4	04/04	08:10:00	3	4	3	40	16	2
5	04/05	09:05:00	1	5	4	15	17	1
6	04/06	10:18:00	1	3	3	16	15	2
7	04/07	11:20:00	0	2	3	26	18	1
8	04/08	12:14:00	1	4	5	30	11	2
9	04/09	13:15:00	2	3	4	16	10	3
10	04/10	14:22:00	0	4	3	20	17	1
11	04/11	15:25:00	1	3	5	31	11	1
12	04/12	16:18:00	1	5	4	11	11	2
13	04/13	17:15:00	1	3	2	31	22	2
14	04/14	18:12:00	2	4	7	24	10	3
15	04/15	19:14:00	1	2	5	29	16	1
16	04/16	20:20:00	1	3	3	17	11	4
17	04/17	07:12:00	1	4	6	17	12	2
18	04/18	08:15:00	1	3	4	34	15	1
19	04/19	09:05:00	1	4	7	32	8	3
20	04/20	10:17:00	2	3	5	17	12	2
21	04/21	11:20:00	0	4	7	30	12	1
22	04/22	12:04:00	0	2	5	25	9	2
23	04/23	13:15:00	1	3	6	22	12	3
24	04/24	14:12:00	1	2	4	14	7	2
25	04/25	15:20:00	1	1	5	17	9	4
26	04/26	08:15:00	1	1	6	20	15	2
27	04/27	09:30:00	1	1	7	27	12	2
28	04/28	10:45:00	0	2	5	20	13	1
29	04/29	11:20:00	0	2	5	12	21	3
30	04/30	12:36:00	1	3	4	18	13	2
			$V(6)$	$V(4)$	$V(3)$	$V(1)$	$V(2)$	$V(5)$
			29	88	139	694	390	60
			$PC(6)$	$PC(4)$	$PC(3)$	$PC(1)$	$PC(2)$	$PC(5)$
			2.07	6.29	9.93	49.57	27.85	4.29

Then the total cost for the whole month for each cause is

$$\text{total cost cause 1} = \$80/\text{occurrence} \times \ \ 29 \text{ occurrences} = \$ \ \ 2,320$$

$$\text{total cost cause 2} = \$85/\text{occurrence} \times \ \ 88 \text{ occurrences} = \$ \ \ 7,480$$

$$\text{total cost cause 3} = \$90/\text{occurrence} \times 139 \text{ occurrences} = \$ 12,510$$

$$\text{total cost cause 4} = \ \ \$3/\text{occurrence} \times 694 \text{ occurrences} = \$ \ \ 2,082$$

$$\text{total cost cause 5} = \ \ \$4/\text{occurrence} \times 390 \text{ occurrences} = \$ \ \ 1,560$$

$$\text{total cost cause 6} = \$30/\text{occurrence} \times \ \ 60 \text{ occurrences} = \$ \ \ 1,800$$

$$\text{Sum} = \$ \ 27,752$$

The percent dollar contribution for each cause is

$$\text{percent contribution cause 1} = \frac{100 \times 2,320 \ \$}{27,752 \ \$} = \ \ 8.36\%$$

$$\text{percent contribution cause 2} = \frac{100 \times 7,480 \ \$}{27,752 \ \$} = 26.95\%$$

$$\text{percent contribution cause 3} = \frac{100 \times 12,510 \ \$}{27,752 \ \$} = 45.08\%$$

$$\text{percent contribution cause 4} = \frac{100 \times 2,082 \ \$}{27,752 \ \$} = \ \ 7.50\%$$

$$\text{percent contribution cause 5} = \frac{100 \times 1,560 \ \$}{27,752} = \ \ 5.62\%$$

$$\text{percent contribution cause 6} = \frac{100 \times 1,800 \ \$}{27,752 \ \$} = \ \ 6.49\%$$

The Pareto diagram for dollar cost for the month of April based on the above calculations is shown in Fig. 14-2.

Further examination of Fig. 14-1 shows that in terms of occurrences, causes 4 and 5 are responsible for 77.42% of the problem. The Pareto diagram of Fig. 14-2, on the other hand, shows that causes 2 and 3 are responsible for 72.03% of the dollar cost. It is apparent from this example that the priority of causes changes depending on the weighting coefficients used.

Thus eliminating causes 2 and 3 reduces cost by 72.03%, while eliminating causes 4 and 5 reduces cost by only 13.12%. Two additional factors, however, need to be taken into account. One is the cost of eliminating each cause and the other is the duration of the process.

Consider, for example, a process that runs continuously for a long period of time. Eliminating causes 4 and 5 may have a two-year payout, while eliminating causes 2 and 3 may have a five-year payout, because they are more costly. However, in absolute terms, over a long time, say 10 years or more, it may still be beneficial to eliminate causes 2 and 3.

Also, it may only be possible to minimize the effect of a given cause as

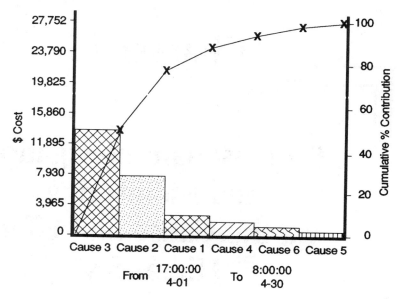

Figure 14-2   Pareto Diagram for Cost.

opposed to eliminating it. However, in accordance with the SQC/SPC philosophy, one strives for continuous improvement.

14.2 USES OF PARETO DIAGRAMS

The example of the previous section shows that the Pareto diagrams for the number of occurrences and costs present different views of the problem. Moreover, they complement each other in terms of indicating the way toward improvement. They are used to prioritize causes of problems and point the way to maximizing the return for the effort expanded.

Chapter 15

Process/Plant Navigation and Monitoring via the Cause and Effect Hierarchy

This chapter presents the cause and effect diagram and its use as an active mechanism for information access analysis and monitoring. Moreover, the concept is expanded to provide a cause and effect hierarchy for process/plant information organization, analysis, and monitoring.

15.1 CAUSE AND EFFECT DIAGRAM

Figure 15-1 shows a cause and effect diagram (CED), also called a fishbone diagram because of its structure. It was developed by Kaoru Ishikawa (reference 9) in his work with quality circles in Japan, and it is sometimes referred to as the Ishikawa diagram. It is used to document and classify the relationships between quality effects and their causes. The head of the fish indicates a specific effect, while the entries on the bones indicate the causes. As originally used, the diagram was drawn by hand on paper, and the causes and effect were also hand written onto the diagram.

For the distributed electronic control and information systems supported by communications networks described in Chapter 1, the cause and effect diagram not only documents the effect and its related causes, but it is also used as an

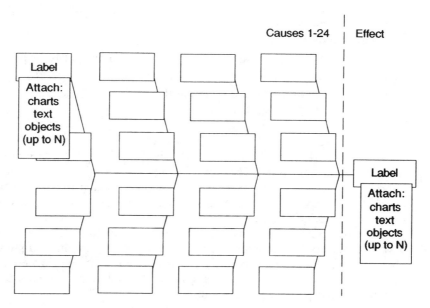

Figure 15-1   Cause and Effect Diagram.

active information access mechanism in a computer-display system. This is achieved by linking charts, text files, and other objects to each cause and effect box. Each cause and effect box is an active screen area selected by touch or other pointing device.

The text objects are text files intended to provide useful information to the operator about a cause variable and its effect, as well as guidance as to what action to take when indicated.

Any of the SPC charts and analysis diagrams described in this book, as well as any other analysis tools that use the real-time database, can be attached to the cause and effect boxes. Other objects that can be attached are process and instrumentation diagrams for units and subunits to which a set of cause and effect variables belongs. All attachable objects are preconfigured and stored in electronic storage devices to be accessed during plant/process operation.

The statistical process control charts described throughout this book define all the information necessary to retrieve collected process data, merge or transform it, subgroup it, perform the statistical calculations appropriate to the chart type, plot and display the results, and apply statistical control rules to the data to determine when the variable is out of statistical control.

The chart display presents the chart as configured and enables the operator or analyst to:

• Check for rule violations and provide indication of not-in-statistical control

- Display embedded chart help
- Display or enter notes
- Display or change chart parameters, either temporarily or permanently
- Display the results of internal chart calculations
- Print the entire screen containing the chart
- Print a standard chart report

Referring to Fig. 15-1, the box labeled "effect" represents the effect, and cause boxes 1 through 24 represent the causes. Each box has a label for identification purposes. To the effect box and to each cause box, one can link up to N chart and/or text and/or other types of objects. When more than one object is linked to the box, the one to be displayed is selected via a menu, which lists the names of the charts and text files.

Figure 15-2 shows an example of a cause and effect diagram with charts and text attached to it. An effect chart can be displayed simultaneously with a cause chart using the window capability of display workstations. Moreover, a number of charts can be displayed simultaneously, depending on screen size and desired resolution, or by using multiple screens.

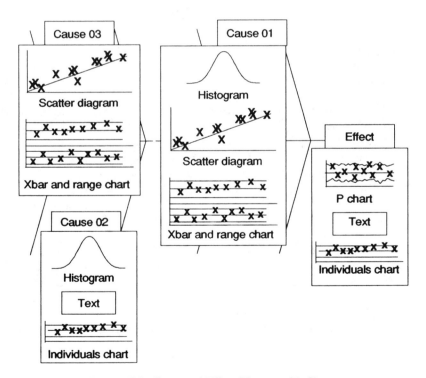

Figure 15-2 Cause and Effect Diagram with Charts.

One can use individual CEDs to organize a plant in a flat kind of structure. That is, each major process unit with its associated subunits is organized using a single, dedicated CED. Another way to organize process or plant information is to use the hierarchical structure described next.

15.2 CAUSE AND EFFECT HIERARCHY

To the effect box of a CED, one can attach up to N objects consisting of charts, text, and other objects. To each of the cause boxes, on the other hand, one can link up to N objects, as in the effect box, or link another CED. If another CED is linked to a cause box, it is displayed automatically when the cause box is activated.

The next CED can access objects or another CED, and so on. Thus the CED is organized into a hierarchical structure. This *cause and effect hierarchy* is then used as an active information access mechanism that can zoom in on more and more detail as one descends the hierarchy. Note that the effect box at any level of the hierarchy can only access objects.

Typically, a plant or process can be divided into functional areas with each area requiring its own *area cause and effect hierarchy* (ACEH) for operation. Areas are made up of units, and units are made up of subunits. Figure 15-3 illustrates an area cause and effect hierarchy. At the top level of the hierarchy, level 1, is the area CED, which shows three cause boxes implemented. Cause 01 accesses the level 2, unit 1 CED, and cause 02 accesses the level 2, unit 2 CED, while cause 03 accesses a chart. When all 24 cause boxes are configured to access CEDs, level 1 can have up to 24 units linked to it.

The level 2, unit 1 CED has three cause boxes implemented: cause 01, accessing a chart; cause 02, accessing the level 3, subunit 1; and cause 03, accessing level 3, subunit 2. When level 3 is the bottom of the hierarchy, as is the case here, the cause boxes at this level only access chart or text objects. However, the hierarchy does not have to stop at level 3. It can continue as necessary down to level N.

To build a plant or process hierarchical structure, the area cause and effect hierarchies are linked together by the plant CED. This means that the plant structure accesses the configuration of the area structures and their associated objects and executes them and displays them on demand. Figure 15-4 illustrates the *plant or process cause and effect hierarchy* (PCEH). It shows the cause boxes of the plant CED accessing the area cause and effect hierarchies.

Area operators use the area cause and effect hierarchies to call on demand cause and effect charts to evaluate process operation and make adjustments for improvement. Plant management and analysts can use the plant or process cause and effect hierarchy on demand for analysis. The analysis may indicate the need to make process improvements or to help the operator make operational adjustments to improve or maintain the required product quality.

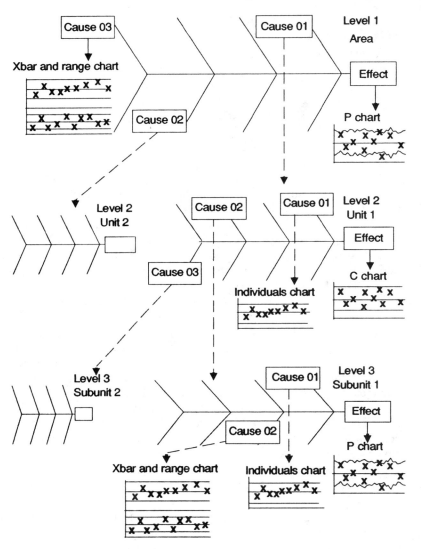

Figure 15-3   Area Cause and Effect Hierarchy.

Recall that every chart attached to the hierarchy is configured to access the real-time database starting at the current time. The current time is the time that the chart is called.

Another use of cause and effect diagrams and hierarchies is for real-time monitoring of a unit, area, or plant, which is the subject of the next section.

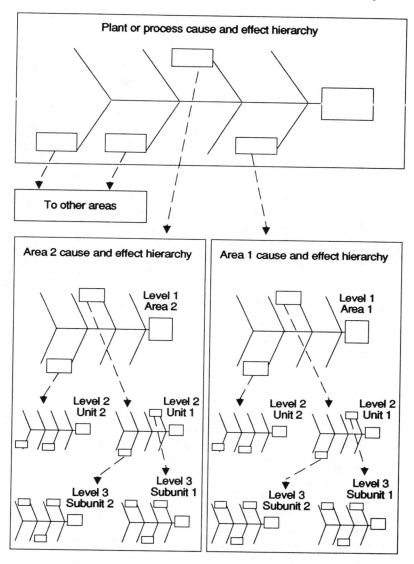

Figure 15-4   Plant or Process Cause and Effect Hierarchy.

15.3 REAL-TIME MONITORING AND THE CAUSE AND EFFECT HIERARCHY

Real-time monitoring of automatically or manually entered samples in the real-time database is implemented by an SPC monitor program, which executes periodically. This program monitors charted variables for out-of-statistical-control conditions. The monitor period for a given chart is based on the chart subgrouping options and the variable's sample period.

At each chart's monitor period, the monitor obtains sufficient data from the database to check the latest subgroup of samples and the necessary number of previous subgroups for rule violations. The number of previous subgroups needed depends on the rules used.

When SPC chart variables are monitored in real time for out-of-statistical-control rule violations, cause and effect diagram boxes linked to monitored charts show the out-of-statistical-control alarm status of the charts via the background color of the box, as follows:

- Red for in-alarm
- Yellow for acknowledged
- Normal color for not-in-alarm

If a given cause or effect box is linked to one or more charts, the collective alarm status of the charts is indicated by the background color of the box.

Alarm acknowledgment is also done by first selecting the CED box-in-alarm and then selecting the chart in-alarm attached to that box. The chart display itself is then used to acknowledge alarms for the displayed chart.

The operator observes the charts for a quality or other type of variable, interprets them, and uses the cause and effect diagrams to track the causes of out-of-statistical-control variables, observes the charts for those variables, and implements the necessary control actions. Typical control actions consist of the following:

- Changing appropriate variable targets
- Retuning the controllers and modifying associated control functions
- Improving control of upstream units to minimize the introduction of systematic variation into the downstream process

Figure 15-5 shows the effect box and cause box 02 in-alarm by changing the background of the boxes as indicated. Note that in the actual situation this would be done by changing the background color of the boxes to red.

Figure 15-6, on the other hand, shows an area cause and effect hierarchy and real-time monitoring. The rule violation on the individuals chart in level 3, subunit 1 is indicated by changing the background color of cause box 01 to red, as indicated. In addition, the alarm status is also shown in each cause and effect diagram box linked to this chart all the way to the top of the hierarchy. Specifically, the individuals chart-in-alarm status is shown by changing the background color in (from the bottom up) level 3, subunit 1, cause box 01; level 2, unit 1, cause box 02; and level 1, area, cause box 01.

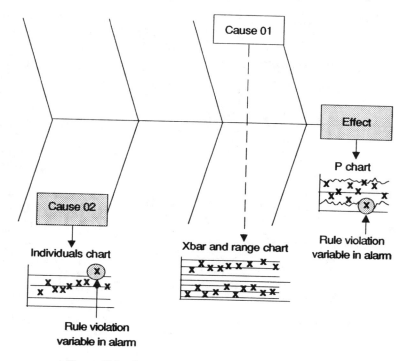

Figure 15-5 Cause and Effect Diagram and Monitoring.

The novelty of the approach presented here is the use of the cause and effect diagram for quality monitoring and information access by attaching SPC charts, text, or other objects to the effect and cause boxes of the diagram and by automatically monitoring the statistical state of the charts. Moreover, by linking CEDs at different levels, one can build area cause and effect hierarchies for plant areas. The area hierarchies can then be linked together to obtain the process or plant hierarchy.

The individual cause and effect diagrams and the area cause and effect hierarchies provide consistent monitoring and information access for the local areas, while the process or plant cause and effect hierarchy provides for consistent information access for the whole plant.

It has been pointed out that many different types of objects can be attached to the cause and effect diagrams and hierarchies. The structure then can be used for information analysis to satisfy the needs of everybody in the plant or process. For example, the operator can use it to call appropriate objects to operate the plant, while the engineer/analyst can use it to call analysis tools to determine required improvements.

A single hierarchy can be used for the whole plant or process. As such, it

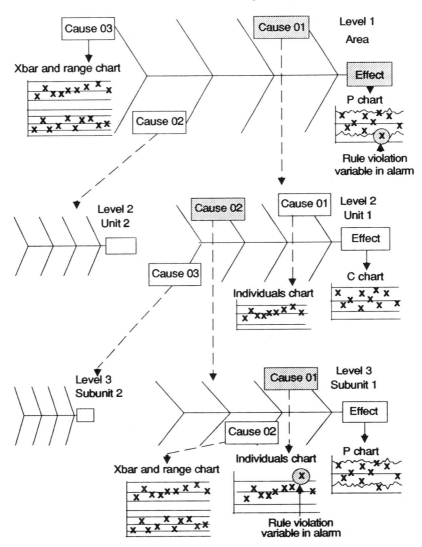

Figure 15-6   Area Cause and Effect Hierarchy and Monitoring.

provides for sharing of the same information by everybody and reinforces the commitment to continuous improvement and increased productivity. The fact that everybody navigates through plant or process information from beginning to end increases the awareness of both internal and external customers and suppliers of the need for quality improvement.

Chapter 16

Functions
of Real-time SPC

This chapter presents the real-time SPC functions required for on-demand analysis and for plant or process operation. Figure 16-1 provides a functional overview of real-time SPC. The main functions are as follows:

- Configuration
- Operational displays
- Monitor and alarm

These functions are supported by functions for the following:

- Data collection and access
- SPC configuration and runtime data and access
- Calculations
- Report generation

16.1 DATA COLLECTION AND ACCESS

Data collection is performed by the real-time database. The meaning of real-time database, as used throughout this book, is covered in detail in Section 1.2 and is repeated here for clarity. The *real-time database,* shown in Fig. 16-1, refers col-

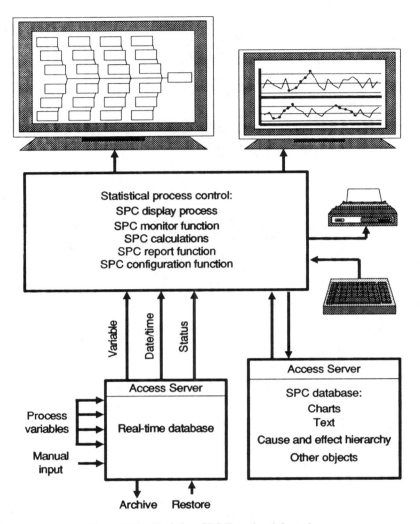

Figure 16-1   Real-time SPC Functional Overview.

lectively to all the distributed databases that collect and store the value, date/time, and status for all plant variables from the current (present) time to some time in the past.

Since the real-time database collects data from the current time to some time backward, it is always tied to the current time. Thus the SPC charts and other analysis tools, when requested, automatically access data from the current time backward. Thus, when a chart is called by plant engineers, chemists, other analysts, or plant operations management for analysis, it starts with the current time.

To provide maximum flexibility for analysis and for operation, collected samples need to be accessed in the following ways:

1. Start date/time, number of subgroups backward in time
2. Start date/time, number of subgroups forward in time
3. Start date/time, time span backward in time
4. Start date/time, time span forward in time

The default start date/time is the current system date/time as maintained by the system real-time clock, and the samples are accessed by method 1 or 3.

16.2 CONFIGURATION AND RUNTIME DATA ACCESS

All access to SPC configuration and runtime data is provided by the SPC database access server, as shown in Fig. 16-1.

16.3 CALCULATIONS

A library of functions provides the mathematical transformations and statistical calculations required for chart displays and monitoring.

16.4 REPORTS

Configuration Reports

A set of predefined configuration reports is provided for each chart type and cause and effect diagram by the SPC report function, on demand via the SPC configuration function. The following configuration report types are provided.

- List of all charts and cause and effect diagrams
- Detailed report of all charts and cause and effect diagrams
- Detailed report for all charts of a given type, for example, all xbar and range charts, or all cause and effect diagrams

Operational Reports

A chart can be called up for display, its parameters changed, and a report generated consisting of tables of raw and calculated values and notes for the selected time period. The PRNTSCRN function in the chart display is used to obtain hard copy of the plotted chart. Operational reports are produced by the report function, on demand via the display function.

16.5 CONFIGURATION FUNCTION

SPC configuration is performed by the configuration function, which interacts with the user via workstation displays and updates the SPC definition files via the SPC database access server. The configurator presents the top menu bar, shown in Fig. 16-2, which contains the following buttons:

HELP

This button displays embedded help for configuration.

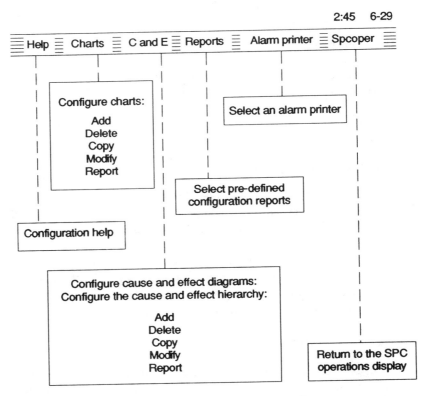

Figure 16-2   SPC Configuration Menu Bar.

CHARTS

This button provides the displays needed to configure charts. It provides the ability to add, delete, copy, modify, and print a chart definition.

C and E

This button provides the displays needed to configure cause and effect diagrams and the cause and effect hierarchy. It provides the ability to add, delete, copy, modify, and print a cause and effect diagram definition.

REPORTS

This button provides the ability to select and print a predefined configuration report.

ALARM PRINTER

This button provides the ability to select an alarm printer.

SPCOPER

This button provides the ability to exit the configuration function and enter the SPC display function.

Chart Configuration

Each chart is configured as a separate, named instance of one of the supported chart types. Chart definitions provide all the information necessary to retrieve collected process data, merge or transform it, subgroup it, perform the calculations appropriate to the chart type, plot and display the results, and check rules for out-of-statistical-control conditions.

The configuration parameters depend on the chart type and are described in detail in the previous chapters pertaining to specific charts. The common functions of chart display and monitoring are described in this chapter.

Cause and Effect Hierarchy Configuration

Each cause and effect diagram is configured as a separate, named definition. The definition specifies the number, position, and title of boxes in the diagram, as well as the charts and text and other objects or other cause and effect diagrams linked to the boxes.

16.6 DISPLAY FUNCTION

SPC provides the following operational displays:

- SPC top-level operations display
- Chart displays
- Cause and effect hierarchy
- SPC current alarm display

SPC Top-level Operations Display

When the SPC display function is activated, it displays the top operations menu bar, shown in Fig. 16-3, which contains the following buttons.

HELP

This button enables the display of embedded operational help.

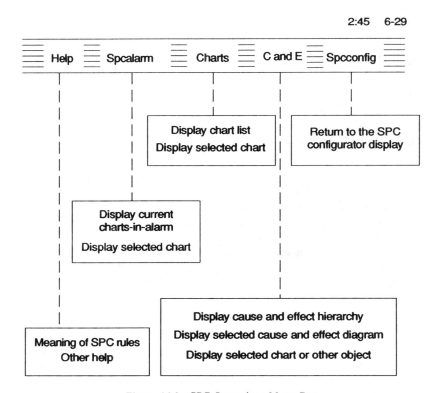

Figure 16-3   SPC Operations Menu Bar.

SPCALARM

This button displays the SPC current alarm display, which contains a list of all charts in alarm. It also contains the acknowledge (ACK) button, which is used to acknowledge alarms. One can then select a chart and acknowledge its alarm or acknowledge all charts in alarm.

When an SPC alarm is detected by the monitor, the SPCALARM button changes color to solid red. If all existing alarms are acknowledged, the button changes color to yellow. When all alarms return to normal, the button returns to its normal color.

CHARTS

This button presents the list of configured charts from which a particular chart can be selected and displayed.

C and E

This button presents the top cause and effect diagram of the cause and effect hierarchy. From there one can select and display a chart, text, or other object or the next-level cause and effect diagram, as described in Chapter 15.

SPCCONFIG

This button provides the ability to exit the display function and enter the SPC configuration function.

Chart Displays

When a chart display is requested, the display function accesses the appropriate chart definition via the SPC database access server, presents the chart as configured, and provides the menu shown in Fig. 16-4, which contains the following buttons:

HELP

This button enables the display of embedded operational help for the selected chart.

SPCALARM, CHARTS, C and E

These buttons are the same as those of the top-level operations display.

2:45   6-29

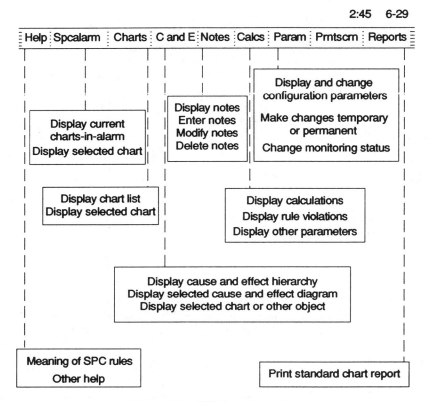

Figure 16-4   SPC Chart Menu Bar.

NOTES

This button enables the display, entry, modification, and deletion of notes associated with a given date/time. Figure 16-5 illustrates this function.

PARAM

This function facilitates on-demand analysis. When the PARAM button is picked, certain chart parameters are displayed. These parameters may be changed and written to the chart definition (on-line configuration change) or just used to re-display the chart. The specific parameters displayed depend on the chart type.

For the xbar and range chart, for example, the parameters that can be changed temporarily or permanently are as follows:

- Transformation type and parameters

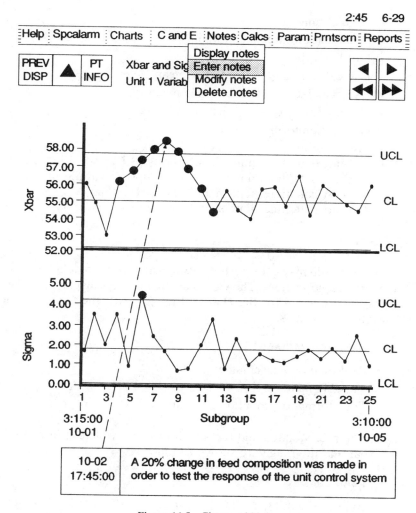

Figure 16-5   Chart and Notes.

- Subgroup type
- Subgroup size
- Time span
- Number of subgroups
- Parameter for $K1$-sigma limits
- Rules for xbar chart
- Rules for range chart
- Official xbar mean

- Official range mean
- Specification targets and limits
- Plot line type
- Use calculated or official xbar mean
- Use calculated or official range mean
- Display specification targets and limits
- On/off monitoring
- Move-in-time:
  1. Start date/time, number of subgroups backward in time
  2. Start date/time, number of subgroups forward in time
  3. Start date/time, time span backward in time
  4. Start date/time, time span forward in time

CALCS

This button displays the results of internal chart calculations as well as other appropriate parameters. The values and parameters displayed depend on the chart type. For example, for the xbar and range chart, the CALCS function displays the following values:

FOR EACH VARIABLE

- Real-time database name
- Variable name
- Transformation equation and parameters

The rule violated and the subgroups that violate it in parentheses for both the xbar and the range are shown as follows:

XBAR CHART

- Rule 1: Freak (8, 13, 14, 19, 20)
- Rule 2: Erratic (19–21)
- Rule 3: Grouping 2 of 3 (7–10, 12–15, 18–21)

RANGE CHART

- Rule 1: Freak (19)

OTHER PARAMETERS

- Subgroup type
- Number of subgroups (configured)
- Number of subgroups (actual)

- Parameter for $K1$-sigma limits
- Upper specification limit xbar (optional)
- Target xbar (optional)
- Lower specification limit xbar (optional)
- Upper specification limit range (optional)
- Target range (optional)
- Lower specification limit range (optional)

CALCULATED AND/OR OFFICIAL VALUES FOR:

- Upper control limit xbar
- Central line xbar
- Lower control limit range
- Upper control limit range
- Central line range
- Lower control limit range
- Xbar mean
- Range mean
- Sigma prime
- Xbar sigma
- Range sigma

Note that the number of actual subgroups may be less than those configured, because the status of some of the values obtained from the database indicates they are no good (bad) and, therefore, they are not used in the calculations. A bad subgroup in the chart, however, is indicated with a purple solid circle at the top of the chart.

PRNTSCRN

The PRNTSCRN button performs a print screen operation, which can be used to obtain hard-copy of the plotted chart, or the chart can optionally output to a named electronic file.

REPORTS

This button provides the menu of the standard chart reports and a list of report printers. When the user chooses a report and a printer, the report function is activated and produces the report. Reports can optionally output to named electronic files.

Charts are configured to retrieve either a given number of variable samples or all available samples for a given time span. If configured for number of samples, the number of samples is computed, using subgroup size (N), number of subgroups (NS), and a subgrouping method, as follows:

$$\text{Size } N: \qquad \text{Number of samples} = NS\ N \qquad (16\text{-}1)$$

$$\text{Size } N \text{ skip } M: \qquad \text{Number of samples} = NS\ (N + M) \qquad (16\text{-}2)$$

If configured for a time span, all values collected in that time span are retrieved, and as many subgroups as possible are formed.

The configured variable transformation (if any) is performed on the samples, and subgroups are formed according to the configured subgrouping method. The appropriate statistical calculations are then performed and the results plotted, using the configured plot line type. Any configured statistical control rule checks are then performed, and the results are indicated on the display itself and can also be listed using the CALCS button.

Figures 16-5 and 16-6 show three different ways of displaying a chart. Figure 16-5 shows the subgroups plotted using small, solid circles connected by a line. Subgroups that violate a rule, indicating that a variable is not in statistical control, are shown with large solid red circles, as indicated by the large solid black circles. Figure 16-6(a) shows the subgroups connected by lines without the circles, but rule violations are again shown in large solid red circles. Finally, Fig. 16-6(b) shows the subgroups indicated by small solid black circles, while rule violations are again shown in large solid red circles. It is recommended that central lines be shown using green lines. It is also recommended that control limits be shown using red lines.

Arrow Buttons

Figure 16-5 also contains the PT INFO, the PREV DISP, and five arrow buttons. Examples of the PT INFO function are shown in Figs. 2-2, 5-1, and 12-1. This function provides detailed information for the selected subgroup. In addition to the value, it could also show the rules violated, the status, and other information deemed appropriate. The operator can select and display the detailed information for any subgroup on the displayed chart. Putting detailed information in the background and making it readily available for display as needed reduces chart display clutter and confusion and makes the charts exceedingly operational.

The PREV DISP button either returns to the parent cause and effect diagram if implemented or to the top-level operations display.

The up-arrow button returns to the previous display. The left and right arrow buttons are used to scroll the charted data in time. Each touch of the single left arrow button moves the charted data backward in time by one subgroup. Each touch of the double left arrow button moves the charted data backward in time by half of the displayed data. The right arrows do the same in the forward direction.

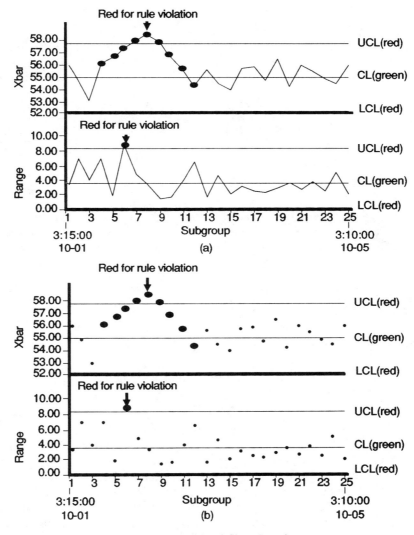

Figure 16-6   Additional Chart Functions.

Cause and Effect Hierarchy Display

Chapter 15 shows examples and provides a detailed explanation for the display of the cause and effect diagrams and the chart and text objects linked to them, as well as the plant cause and effect hierarchy.

Current Alarm Display

The SPC current alarm display shows each record in the SPC current alarm file, one line per record, in chronological order with the most recent violation time first. Each line contains the following:

- Chart name
- Name and number of rules violated
- Violation date and time
- Acknowledgment date and time
- Return-to-normal date and time

The line is red in background color when the chart is in alarm, yellow if the alarm is acknowledged, and normal color if the alarm has returned to normal. The alarm can be acknowledged in this display, and the chart named in each line can be selected for display.

16.7 MONITOR FUNCTION

Charts of the following type may be chosen for real-time monitoring and alarming:

- Individuals, xbar and range, xbar and sigma, CUSUM
- *P, NP, C, U*

With the exception of CUSUM charts, which use the configured decision intervals $h1$ and $h2$, each monitored chart should use up to three rules chosen from the 11 rules listed in Table 2-3. Choosing a maximum of three rules for real-time monitoring and operation is reasonable, because choosing too many rules would create confusion in a plant operating environment. Of course, the plant analyst can enable all rules for plant analysis. However, one should be aware that plant operators cannot be expected to have a high level of analytical skills. Therefore, it is imperative to analyze the variable over a long period of time and use two or three rules that are most appropriate for what has happened in the past as well as what the objectives are for the future.

The monitoring of variables collected by the real-time database is implemented by the SPC monitor process, which executes periodically. This process monitors charted variables for out-of-statistical-control conditions, prints alarm messages, and updates the current alarm file.

At each chart's monitor period, the monitor obtains sufficient data from the real-time database to check the latest subgroup of samples, as well as many earlier subgroups, as needed for rule violations. When the alarm status changes, an alarm message is printed and the SPC current alarm file is updated via the SPC database access server.

16.8 WINDOWS AND CAUSE AND EFFECT ANALYSIS

When a cause and effect diagram is displayed, one can select a cause or effect box and display any of the attached objects to them. However, the windowing capability of workstations and/or multiscreen display capability is used to simultaneously display charts for effects and causes. This facilitates cause and effect analysis by the operator, the engineer/analyst, and operations management.

Chapter 17

Process/Plant Analysis and Operational Setup

The main purpose of this chapter is to provide the necessary analysis steps for selecting the appropriate tools and organizing a process or plant for analysis and operation. Chapter 18 discusses the real-time plant analysis and operation itself.

As described in Section 1.3, the term real-time database refers collectively to all the distributed databases that collect and store the value, date/time, and status for all plant variables from the current (present) time to some time in the past. Thus the SPC charts and other analysis tools, when requested, automatically access data from the current time backward.

It was also pointed out that the on-line storage for the whole distributed real-time database should be months or years, to provide for ready process analysis, operator training, reporting, and other functions. The on-line storage combined with archive storage on electronic media should contain the real-time database for the lifetime of the process or plant. Of course, all or parts of the archived real-time database should be readily restored for on-demand analysis and be accessible by the same SPC charts and tools that access the on-line stored real-time database.

According to Section 1.3, the real-time database contains cause, effect, and attribute variables and causal relationships. Therefore, the tools of SPC can be classified according to variable type and analysis done as follows:

1. SPC tools for cause and effect variables:
   a. For analysis and monitoring of individual samples:
      - Individuals histogram
      - Individuals chart
      - Scatter diagram for autocorrelation
   b. For analysis and monitoring of subgrouped samples:
      - Xbar histogram
      - Xbar and range/sigma chart
      - CUSUM chart
2. SPC tools for attribute variables
   a. For monitoring of fraction and number defective:
      - P chart
      - NP chart
   b. For monitoring of defects and defects per unit:
      - C chart
      - U chart
3. SPC tools for cause and effect analysis
   - Scatter Diagram for Cross-Correlation
   - Pareto Diagram
   - Cause and Effect Diagram
4. SPC tools for closed-loop control
   - CUSUM controller
   - Optimum Setpoint Controller

The real-time distributed database collects and stores all the required samples and their attributes. Therefore, the SPC database for any charts associated with a particular variable contains only the chart definitions. It does not need to store any raw or calculated values. The definitions take very little electronic storage space. Therefore, storage space is not a consideration in configuring multiple charts per variable as deemed necessary.

Using the preceding classification of tools, the main steps for operational setup are as follows:

1. Information organization and cause and effect analysis
2. Analysis and chart selection for cause and effect variables
3. Analysis and chart selection for attribute variables
4. Use of the Pareto diagram
5. Use of the SPC closed-loop controllers
6. Use of charts for long-term trending and partnerships
7. Other considerations in using SPC tools

17.1 INFORMATION ORGANIZATION AND CAUSE AND EFFECT ANALYSIS

The process or plant can be divided into areas, units, and subunits. The cause and effect hierarchy described in Chapter 15 is used as an active information mechanism to organize and access plant charts, text, and other objects. The basic organizing structure is the cause and effect diagram shown in Fig. 15-1. Figure 15-3 shows an area cause and effect hierarchy, and Fig. 15-4 shows the plant or process cause and effect hierarchy.

Consider a continuous or semicontinuous plant or process unit and its cause and effect variables. The effect variables are typically product flows and compositions and energy used per unit of product. The cause variables are usually feed flows and compositions, catalyst flow when used, pressure, temperature, and residence time of the unit.

These are the typical cause and effect variables of intermediate process units. However, the final products even in continuous plants, such as refineries and pulp and paper mills, may be shipped in containers, such as barrels or tank cars or rolls of paper; then one of the effect variables can be fraction or number defective.

The cause and effect diagram shown in Fig. 15-1 has 24 cause boxes and one effect box. Up to N charts and other objects can be attached to each box. It is recommended that each cause box be associated with a single variable with its charts and text to avoid confusion. The effect box can contain all necessary variables associated with the product or products.

A process unit is designed to make products, and much is known about cause and effect based on chemistry and/or physics and other sciences or experience gained in the laboratory or pilot plant. However, once the unit is started and operated for a period of time and data are collected in the real-time database, the scatter diagram can be used to determine cause and effect quantitatively. The diagram provides a visual display, as well as the cross-correlation value for each pair of cause and effect variables.

In many cases the effect variable is directly related to the ratio of two cause variables. For example, the melt index of a polymer is related to the ratio of catalyst to feed, or pulp brightness is related to the chemical-to-pulp ratio fed into the bleaching stage. There are many other examples of important ratio variables throughout the process industries.

The scatter diagram can also display and compute the cross-correlation of an effect variable versus the ratio of two cause variables or the ratio of two effect variables versus the ratio of two cause variables. The optional time shift compensates for the dynamics of the unit, because changes in the cause variables take time before they start to change the effect variables.

Once the cause and effect hierarchy has been defined and the cause and effect variables associated with it have been identified for the whole process or plant, the analysis and selection of charts for each variable begin.

17.2 ANALYSIS AND CHART SELECTION FOR CAUSE AND EFFECT VARIABLES

To focus the discussion, consider a single cause variable or a single ratio of two variables and determine which charts to use and why. The first decision is whether to use charts for individual or subgrouped samples. Use charts for subgrouped samples except in cases where the sample periods are too long, say, once every 4 hours or longer.

The following steps describe the sequence of analysis for each variable or each ratio of variables using subgrouped samples.

Step 1. Dynamics and Subgroup Size Determination

Section 4.1 describes how to form subgroups using estimates of process dynamics using Eq. (4-2). This provides the value of the subgroup size N.

Step 2. Autocorrelation and Skip Size Determination

Section 4.2 describes how to compute the autocorrelation of a variable with time shift and Section 4.3 shows how to determine the skip size to minimize autocorrelation. This provides the value of the skip size M.

Step 3. Xbar Histogram and Normality Measures

Subgroup type N skip M, with the values of N and M determined in steps 1 and 2, respectively, is then used to compute the xbar values for a sufficient number of subgroups to build a reliable xbar histogram. The analyst then recomputes and redisplays the xbar histogram by moving-in-time through the real-time database in order to check the normality of the xbar values of the data, using visual inspection and the rule of thumb given by Eqs. (2-8) and (2-9).

Note that normality is *not* necessary in applying the rules of Table 2-3. However, the closer the distribution is to the normal, the easier it is to determine that the probability of a rule violation is approximately 0.3%, as described in Section 2.6.

Step 4. Charts, Rules, and Text Objects to Use for Operation

The recommended charts for operation are the xbar and sigma and the CUSUM charts with the N skip M subgrouping method, with the N and M values as determined above. The reasons for this choice are given in Section 2.6 and, as described there, the recommended rules for the xbar and sigma charts are any combination of the first three rules of Table 2-3.

The two charts are used in a complementary fashion. The xbar and sigma chart with any of the three recommended rules provides faster response for large variations. The CUSUM chart, on the other hand, provides fast response for sustained small deviations from the target. At the same time, the CUSUM chart provides acceptable response for some of the other rules of Table 2-3, such as rules 5 through 11.

At this point, a text file is created that contains an explanation for the operator of the cause variable under analysis and its effect. Moreover, the text file provides operator guidance as to what action to take when indicated.

Thus for each variable the recommended objects are as follows:

- Xbar and sigma chart
- CUSUM chart
- Scatter diagram for autocorrelation
- xbar histogram
- Text

Summarizing, the xbar and sigma and CUSUM charts are used to monitor the variable. The scatter diagram is used to monitor its autocorrelation, and the xbar histogram is used to monitor normality. The text is used for explanation and guidance.

Usually, the preceding four steps are sufficient to define the charts for each variable or ratio of two variables.

Step 5. Mathematical Transformation When and Why

Section 2.5 discusses the use of mathematical transformations. There are times when the mathematical transformation of a variable is useful. This should only be done after careful analysis to ascertain that the mathematical transformation does not obscure other problems, such as autocorrelation.

If the distribution of xbar values using the M skip N subgrouping method is not acceptably normal, then try the mathematical transformation of Table 2-2 to determine if the transformed variable will make the distribution of the xbar values acceptable. If it does, then use the transformed variable for the xbar and sigma, the CUSUM, and the xbar histogram in step 4.

Make sure that the analysis is done for sufficient data for the variable by moving-in-time through its real-time database to ascertain the need for transformation.

17.3 ANALYSIS AND CHART SELECTION FOR ATTRIBUTE VARIABLES

The SPC tools for monitoring fraction and number defective and defects and defects per unit are the following:

- *P* chart
- *NP* chart
- *C* chart
- *U* chart

A plant or process consists of many areas, units, and subunits in combinations of serial and parallel operations, starting with the raw materials to the final products. The raw materials can be truly totally unprocessed or they can be products from other plants or processes. From raw materials to final products, there can be many intermediate products, each of which uses the SPC tools listed above.

The raw materials, when they are products of another plant, and the final products of the plant of interest, in addition to the tools listed above, can also use the attribute variable charts.

Consider, for example, rolls of paper being the final product in a pulp and paper mill. In this case, the *P* or *NP* chart can be used to monitor the fraction of defective rolls or number of defective rolls, respectively. In addition, one can take a 1-meter sample of paper from a roll and count the number of defects and build a *C* chart or a *U* chart if the sample size changes.

These charts are used to monitor final product. Typically, rule 1 of Table 2-3 is the one to use, although under certain circumstances explained in Chapters 12 and 13, all eleven rules can be used. When these charts indicate that the variable is not in statistical control, then the problem can be determined and action can be taken by using the quality and causal variable charts for the areas, units, and subunits in the plant cause and effect hierarchy.

The customers that buy the paper rolls may require from their suppliers some of the attribute charts described above. In addition, they may require the xbar and sigma chart for the paper strength, basis weight, and brightness for a given period of time, say a week or a month. All this information is necessary to develop the long-term confidence and partnerships between customers and suppliers required to maintain and improve quality and productivity. For every plant or process, there are suppliers and customers. Therefore, a given plant must require the same type of information and partnership from its suppliers that it provides to its customers.

There are process plants, however, where the products flow continuously through pipes to their customers. For example, a refinery provides ethane feed to an ethylene plant, which in turn provides ethylene to a polyethylene plant, which is used to make plastics. For plants of this type, the supplier/customer partnership is based on SPC charts for the quality of the products. For example, one can use the xbar and sigma chart for ethane composition in the refinery product, or the ethylene composition in the ethylene plant product, or the melt index of the polyethylene in the polyethylene plant product. Of course, the plant that receives polyethylene and makes plastic bags can use a P chart for the fraction of defective bags.

17.4 USE OF THE PARETO DIAGRAM

The Pareto diagram determines the 20% of the causes of product rejection that result in 80% of the problem. Thus, in addition to tracking the fraction defective of the rolls of paper from the paper mill, one can also track the causes (reasons) for rejection. Over a period of time, say a month, one can build a Pareto chart as described in Chapter 14 to determine the 20% of the causes that result in 80% of the problem.

Therefore, the Pareto diagram focuses the order of priority of projects to be undertaken to get the maximum return for the investment expended. For example, investing to solve the 20% of the causes determined above results in 80% improvement.

As is the case with all other SPC tools, Pareto diagrams can also be attached to the plant cause and effect hierarchy. This way one can build a Pareto diagram for any length of time, say a day, week, month, or year, and move-in-time through the real-time database of the variables involved.

As described in Chapter 14, the weighting coefficients should be used to build Pareto diagrams to provide different ways of looking at the data. For example, consider the case where the cost of each occurrence for a cause of rejection is very small relative to the other causes. Assume that this cause has occurred the largest number of times, which results in its being the largest in the number of occurrences Pareto diagram. Now, building a Pareto diagram for cost, the same cause is the smallest. Therefore, it is recommended to build Pareto diagrams for the same causes to provide different views using the weighting coefficients.

Process control systems contain a number of alarms for their variables, for example, high-high, low-low, high and low absolute alarms, and deviation alarms. Moreover, the alarms are prioritized: priority 1, 2, . . . , and so on. The alarms for a process unit can be collected by the real-time database, and a Pareto diagram can be built for all alarms of a particular priority. The Pareto diagram can show which alarm occurred for the longest time over a desired time span, say a shift, day, or week. Pareto diagrams can also be attached to the plant cause and effect hierarchy for real-time access and alarm analysis.

17.5 USE OF THE SPC CLOSED-LOOP CONTROLLERS

The SPC tools for closed-loop control are the

- CUSUM Controller
- Optimum Setpoint Controller

Both of these controllers use subgrouped samples and manipulate downstream variables to maintain a controlled variable at a desired target or set point. They are described in detail in Chapters 10 and 11. They have unique features and, when used appropriately with other traditional process control tools, they provide more effective overall closed-loop control.

17.6 USE OF CHARTS FOR LONG-TERM TRENDING AND PARTNERSHIPS

The charts can also be used for long-term trending of plant operation, as described in Section 5.6. This is achieved using charts for subgrouped samples. These charts provide a longer-term view of process variables, as opposed to standard trend charts, which are similar to the individuals chart.

Of course, histograms can always be used to monitor the distribution of samples, as well as for process capability analysis. Also, the scatter diagram can be used to determine the autocorrelation of a single variable or the cross-correlation, that is, cause and effect relationship, between two or more variables.

Charts can also be used to establish long-term relationships between customers and suppliers, as described in Section 17.3.

17.7 OTHER CONSIDERATIONS IN USING SPC TOOLS

The plant cause and effect diagram provides the basis for approaching the plant as a whole. It considers each of the real products or waste of the facility and their individual or collective market issues or problems. Problems should first be identified by using SPC charts for quality variables and specific measures of performance and tracing them to their sources. To do this requires the establishment and evaluation of intermediate measures of performance to arrive at problem solutions.

Reaching statistical control with inadequate traditional basic and advanced regulatory control of time-dependent correlated variables is difficult. However, the SPC tools are very helpful in focusing the attack on traditional control problems. The plant-wide approach leads naturally to a unit approach.

The unit approach uses cause and effect diagrams for the individual units. It involves identifying and charting quality variables and measures of performance of the raw materials and products of the unit. To do this requires the identification of internal as well as external customers and the specifications for the products of each unit.

As stated previously, this approach is similar to the approach followed in the process industry for years. The differences, however, are as follows:

• Emphasizing quality and performance at all levels.

- Providing the necessary tools, knowledge, and understanding down to the level where the product is made and the process is adjusted.
- Using the appropriate SPC tools and traditional process control to deal with the process and its problems.
- Using real measures of performance to make decisions about the process.
- Emphasizing cause and effect all the way from raw materials to final products.
- Using the SPC tools to involve everybody in striving to improve product quality and performance continuously for the individual units and the whole plant or process.
- Using the SPC tools and particularly the plant or process cause and effect hierarchy to make everybody aware of internal as well as external customers and suppliers.

Chapter 18

Process/Plant Operation with Real-time SPC

Plants and processes are divided into areas, units, and subunits and are distributed over a geographic area. They are monitored and controlled by distributed management and control systems that are connected by the communications network. The distributed databases collect and store all measurements from the control and measurement functions, as well as all manually entered data, continuously. This is done day in and day out with the real-time clock providing the date/time for each data value entered. This real-time database is then available for process analysis and operation.

Chapter 17 described how the engineer, analyst, or plant operations management selects the SPC tools to be used for plant operation. The cause and effect hierarchy is used to organize the plant into areas, units, and subunits, with the associated cause and effect variables. Of course, every level in the hierarchy does not have to be used for every plant or process. The number of levels used depends on the plant, the number of variables, geographic distribution, and operations objectives. In many cases dividing the plant into major units is sufficient.

Typically, in a plant an operator is responsible for certain plant areas and units and this operator takes the necessary actions based on the SPC monitoring tools. However, it is important to make available to every operator information about all areas and units that affect and are affected by the operator's actions. In general, the author believes that the more information that is available to all plant personnel, the higher the learning and commitment to quality and productivity. This way the operator's commitment is not limited only to areas of responsibility, but extends to the whole plant.

The SPC tools provided to the operator can be used in two ways. One is on demand; that is, the operator at his or her discretion selects an area or unit cause and effect diagram and displays charts to observe operation and take action. The other consists of scheduling charts for real-time monitoring and, when an alarm occurs indicating rule violation, the operator selects the chart, acknowledges the alarm, and optionally selects other charts and takes action. In either case, the selected chart always displays data starting at the current time, as determined by the system real-time clock, and going back in time through the necessary samples or time span configured.

18.1 ON-DEMAND SPC FOR THE OPERATOR

Section 17.1 described which SPC tools to use for analysis and operation. The cause and effect hierarchy is the recommended way of operating the plant. It is suggested that a single variable be attached to each cause box, while the effect box can have as many as necessary.

For the engineer, analyst, and plant operations management, a number of charts are recommended for a single variable. For example, a scatter diagram for autocorrelation, an xbar histogram for distribution and normality, and an xbar and sigma chart for operation. In addition, all 11 rules are checked to determine which to use during operation.

For operation, however, a maximum of two charts should be used per variable with a maximum of three rules. In addition, text information attached to the same box as the chart should be used to provide explanation and guidance for the operator.

Charts, cause and effect diagrams, and the cause and effect hierarchy are configured to obtain data from the current date/time backward. Then the operator can access the charts in order of preference as follows:

1. By selecting the cause and effect diagram to which the chart is attached and then selecting the desired chart for display
2. By selecting the list of configured charts and then selecting the desired chart for display

The operator observes the charts for the effect variables, interprets them, uses the cause and effect diagrams to track out-of-statistical-control cause variables, and then observes the charts for those variables and implements the necessary control actions. Typical operator control actions consist of the following:

• Changing appropriate set points or targets

- Retuning the controllers
- Improving control of upstream units to minimize the introduction of systematic variation into the downstream process

The operator uses the NOTES function in the chart display to record causes for out-of-control conditions for specific subgroups, as well as the action taken. This provides historical documentation that can be used to improve process understanding as well as for training.

There are times when the operator has done everything possible to minimize variation due to systematic load upsets, but the product still does not meet customer specification. In this case, operations management uses the SPC tools for analysis as described in Chapter 17 to decide on and implement additional actions to improve operation. These actions may consist of the following:

- If the unit uses raw materials from another vendor, ask for charts to prove that they come from processes that are under statistical control and provide product on specification on a continuous basis.
- Use more sophisticated measurement sensors and control functions.
- Modify or replace process units and other equipment.
- Modify or change processing steps.

18.2 REAL-TIME MONITORING AND ALARMING

Section 16.7 describes in detail the monitor function, while Section 15.3 describes the real-time monitoring and the cause and effect hierarchy. As mentioned, charts of the following type may be chosen for real-time monitoring and alarming:

- Individuals, xbar and range, xbar and sigma, CUSUM
- *P, NP, C, U*

In summary, the monitoring of real-time variables is implemented using the monitor function, which executes periodically. This function monitors charted variables for out-of-statistical-control conditions, and when a rule is violated, it indicates it to the operator automatically.

A chart used for monitoring should also be configured for display so that the operator can select it for analysis and observation and can determine what action to take, if any. Red background color indicates the chart is in-alarm, yellow indicates the alarm is acknowledged, and normal color indicates the alarm has returned to normal.

The chart-in-alarm can be acknowledged two different ways, listed in order of preference:

1. By selecting the cause and effect diagram box-in-alarm and then selecting the chart-in-alarm attached to that box.
2. By selecting the current alarm display that contains the list of charts-in-alarm and then selecting the desired chart-in-alarm for display.

Again, the maximum number of rules recommended for real-time monitoring is three. Moreover, the same three rules should also be implemented for on-demand analysis by the operator to achieve consistent analysis and operation.

The real-time monitoring and alarming described above pertain to any variable. The following sections focus on monitoring for specific variables. Section 5.4 discussed monitoring for variables that are not under closed-loop automatic control. Section 5.5 described how to monitor the range of controllability for variables that are under closed-loop automatic control. The next section presents how to monitor model-predicted variables.

18.3 REAL-TIME MONITORING OF MODEL-PREDICTED VARIABLES

For many process units, mathematical models, both steady state and dynamic, are used to predict unit effect (output) variables based on unit cause (input) variables. The models are used for real-time monitoring and/or control of the unit. Typically, the actual effect variables are analyzed in the laboratory. In this case, SPC charts can be used to monitor the error, which is the model-predicted value of the variable minus the actual value obtained in the laboratory.

The monitoring is done in a fashion similar to that for the controlled variable described in Section 5.5. In particular, when the error term varies enough to trigger an alarm, it means that the model no longer predicts well enough for its intended purpose and it should be updated to improve prediction.

18.4 REAL-TIME MONITORING USING EFFECT VARIABLES ONLY

When the cause and effect hierarchy is used for real-time monitoring, as shown in Fig. 15-6, one can use only charts for the effect variables for real-time monitoring. When the effect variable chart is in alarm, the operator can select the appropriate cause and effect diagram in the hierarchy by selecting the effect box-in-alarm and displaying the chart-in-alarm. The operator can then select any of the cause variable charts and display them to determine the problem and decide what action to take, if any. This type of operation is more conservative in the sense that many cause variable alarms are eliminated, but rule violations for them can be requested as soon as the effect variable goes into alarm.

Chapter 19

SQC/SPC
Management Philosophy
and Problem Solving

The purpose of this chapter is to provide a summary of the management philosophy for SQC/SPC that was first introduced in reference 10.

New worldwide demands for quality are being felt and acted on with ever increasing emphasis. For years U.S. industry has emphasized productivity and efficiency. This emphasis, however, has undermined the quality of goods and services and has diminished American effectiveness in the world marketplace. The challenge of providing improved quality can be met by adopting a new approach to quality control. Statistical quality control (SQC) is beginning to provide this approach in the United States and the rest of the world, as it already has in Japan.

Applying SQC during production, as opposed to testing the quality of the end product, has had a startling effect on competitive position and corporate performance, so much so that this practice has spawned a new adage, "Quality cannot be tested into a product, it must be built into the product." Indeed successful practioners of SQC have found that emphasizing quality has yielded rewards far greater than the corresponding emphasis on productivity. The benefits of employing SQC are less rerun, less waste of material, increased capacity, elimination of testing and inspection, and improved customer satisfaction.

19.1 QUALITY IMPROVEMENT

How does a company adopt and implement SQC? Many believe that the solutions to today's quality control problems start with W. Edwards Deming's first point: "Create a constancy of purpose" toward improvement of product and service, with the aim to become competitive, to stay in business, and to provide jobs, which is detailed in reference 1. Constancy of purpose implies a long-range determination to survive. It affects how resources are allocated, how innovation is stimulated, how capital investment is maintained, and how people are developed.

The solutions to today's problems also start with the simple formula for quality improvement given in reference 11.

$$QI = PM + ST + PS \qquad (19-1)$$

where

$$QI = \text{quality improvement}$$
$$PM = \text{participative management}$$
$$ST = \text{statistical tools}$$
$$PAS = \text{problem solving}$$

This formula illustrates that quality improvement requires not only the application of statistical tools and the solutions to problems that arise, but also a committed and involved management. A participative management sets clear goals for the future and actively gets involved in achieving those goals. In this light, SQC becomes more than a mere measure of quality; it is a method for improving quality. It is a management philosophy defining a company's way of doing business.

There are a number of ways to identify a company that is engaged in SQC. If quality groups are actively engaged in improving performance and quality in *all* parts of the business, it is a sign of involvement. If they are qualifying the vendors that supply raw materials to their facilities and are engaged in corresponding activity with their own customers, there is strong evidence of SQC activity. If those directly involved in making the product or delivery of services are charting data related to their production, then they are using SPC and that is a sure sign that they are involved in SQC.

One primary aspect of SQC is statistical process control, which is the principal statistical application tool of SQC. Figure 19-1 represents a definition of the SPC process. It begins with a question: "Is this variable in statistical control?" If the variable is not in statistical control, then action must be taken to assign causes and make the required adjustments to get the variable and finally the process into statistical control.

If the variable is in control, then one asks: "Is the operation satisfactory?" If the quality relative to financial goals is not satisfactory, then an alternative operation is sought. The change in operation could be as simple as a change in

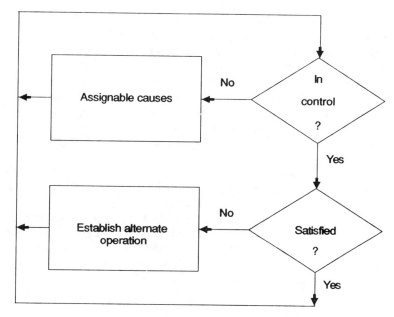

Figure 19-1 The SPC Process.

set points or as complex as plant equipment changes. If the quality is satisfactory, then keep on charting in order to deal with unpredicted changes in operation or pressure from the outside. The very nature of the SPC process solves one of the classic problems confronting industry: "Who is responsible?" In Fig. 19-1, the top loop is the operator loop. The bottom loop is the management loop. This illustrates the difference between working with the system of production versus working on the system of production. What bridges the gap between the two loops is valid statistical data. In both loops, statistical tools not only aid in detecting improvement opportunity, but they also aid in finding causes, devising solutions, and demonstrating economic incentives.

19.2 PARTICIPATIVE MANAGEMENT

Quality gurus insist that the benefits of SQC rely heavily on participative management. Although the tools and techniques for quality improvement have existed for years, it is only recently that U.S. industry has awakened to the urgent need to put them to use. The United States has encountered competition that enjoys an overwhelming quality/price advantage. That competition has spent years conditioning its organizations to the merits of SQC. The United States now must work diligently to catch up.

Some international competitors have set up facilities within the United States, employed U.S. workers, and achieved high quality and efficiency. These organizations generally use *statistical quality control* in support of a management philosophy consistent with constantly improving quality.

What do statistical methods provide that, according to reference 3, generates wholesale changes in management philosophies? The answer lies in the ability of the statistical method to provide information relating to both quality and productivity. The proper use of information requires a level of cooperation between management and other employees hitherto absent from most established business cultures.

A basic doctrine of SPC involves a delineation of "working with the system" and "working on the system." The disciplines of SPC support a clear distinction between the worker's role and management's effort to improve the system. Thus SPC can be used to resolve adversarial issues that too often divide labor and management and weaken the organization's collective competitive impact.

Participative management is both the starting point and the future hope for success in today's quality-conscious business environment. A company that uses participative management will display active

- Leadership
- Teamwork
- Partnership

Leadership

Effective leadership must respond to an ambitious agenda. It must set directions and contribute to the realization of goals by stimulating teamwork among employees. It must work directly with customers to identify product features and quality parameters. It must involve itself with vendors of the materials and technology that form the essence of the production process.

To understand the new requirements for leadership, one must reexamine traditional attitudes composed of certain subliminal values and buried in common practice. For example, the tenet that "the objective of management is to provide leadership, not supervision" from reference 1, flies in the face of authoritarian management techniques. One style relies on inspiration; the other on intimidation, however subtle.

In a company with strong leadership, corporate goals are communicated clearly to all levels. This requires reinterpretation of the goals in terms of job activities at each level within the organization. As contradictions surface between goals and job activities, they must be resolved. Otherwise, the working individual is robbed of the job satisfaction that stimulates quality performance. Leadership should not address its energies to the futile task of identifying and recording failures, but rather remove the *causes* of failure.

SQC underscores the need to eliminate arbitrary numerical goals, although numbers are permissible as an instrument of performance verification. Management's role is to exert the kind of leadership that fits individuals to the job and encourages them to perform their job more effectively. Statistical tools give the individual the capability to evaluate his or her own performance. This leads to a better work climate, where each member does his or her part to deliver a quality product that is both a source of pride and a contribution to the company's product objectives. Team performance can then be rewarded, without regard to specific quotas and numerical goals, because this philosophy achieves them implicitly.

Involved management must perceive firsthand ways to remove the barriers to the resolution of quality problems. Management must stimulate ways to identify cause and effect relationships and resolve problems that compromise quality and performance. In the typical instance, traditional management too often kills the messenger bearing bad news, confusing genuine concern for disloyalty.

Teamwork

Corporate goals require contributions from groups representing differing disciplines and backgrounds. To achieve these goals, it is necessary to stimulate individual contribution within group environments.

The makeup of the team must be consistent with its duties. Corporate organizations call for people to be assigned to functional groups, often by discipline. The team is composed of people from various organizations, and departmental barriers often interfere with the team's efficiency. An involved management is better able to fit the people to the tasks, since these managers are in close contact with the activities and get to know their people on a direct basis.

Teamwork can flourish where fear is replaced by a wholesome concern for personal performance and an attitude of cooperation fuels the group's activities. Fear runs rampant where blame is assessed. Individual pride is greatly enhanced by the use of appropriate tools, processes, methods, and organization.

Partnership

The *cost of quality* is the toll that the customer absorbs for the quality of the product that she or he uses. The lower the cost is the better. Factors related to materials, workmanship, and fit to function are part of that cost. One way to work with cost of quality, whether it be related to the product or raw materials, is to adopt a partnership program with suppliers and customers.

Many customers now qualify their vendors. In so doing, they can establish viable and enduring single-source relationships. This practice springs from the realization that price is not meaningful without some expression of quality. Quality in a vendor–user relationship is made up of many factors, including the following:

Lowest total cost

Simplified accounting and paperwork

Low investment in inventory

Delivery integrity

High conformity to and low variance from target specifications

High applicability

Confidence in vendor's facility

Evidence of statistical control

A sound vendor-customer partnership requires intense cooperation. Much rides on the relationship. One practice common to such relationships is the evaluation audit. This audit is carried out against established criteria derived from the preceding factors. The partnership program involves much more intensive collaboration between companies than has ever been customary within the United States. Many attitudes must be changed to make this kind of relationship happen.

SQC/SPC is not just a passing fad. The theoretical basis has been around for more than 60 years, while the business management principles have been developed over the past 40 years. Statistical quality control has enabled one country, Japan, and many businesses to obtain international trade dominance. The current acceptance by the United States and world industry of SQC techniques and management philosophy will have a significant effect on the process control business. It is expected that these participative management techniques will be employed more and more in the process industries and will become intimately coupled with what is known as process control.

19.3 STATISTICAL TOOLS

The previous chapters describe the tools of statistical process control and their usage. The charts are the visible portion of the tools. The intangible portion is their interpretation, for practice and a firm grounding in causality are needed to achieve excellence in their application. Interpretation will be covered at greater length in the next section.

The statistical tools presented here are precisely those used in Japan to attain the quality and productivity that have made Japan a formidable competitor in many product areas throughout the world. These tools are increasingly applied in the United States at all levels of business operations, with emphasis on building quality into products. The information that statistical tools provide and the way it is presented is the foundation upon which to base a sustained drive toward high quality. Productivity will follow.

Initially, these tools were developed and used in discrete manufacturing. However, they are, indeed, applicable to the process industries. Most of the required data in these industries is already available through on-line transmitters and analyzers and through laboratory analysis entered into information storage and display devices. The appropriate statistical process control charts must be constructed by taking into account process dynamics.

The data patterns on the statistical process control charts reveal what may be the underlying causes of quality variation. Chart interpretation, then, and the use of cause and effect analysis are required in order to solve problems. Providing these tools to operators and operations management and their routine use by them are the basis for continuously improving quality and increasing productivity.

19.4 PROBLEM SOLVING

Statistical process control is unique in its ability to identify problems in any business operation. SPC also helps to classify whether the causes of quality problems are assignable, that is, attributable to a local operation, or whether they are common, that is, attributable to the total operation. Before an identified problem can be corrected, however, a workable solution must be devised and its cost and benefit must be assessed. Three common difficulties confront the investigator of such problems:

1. Experimental error and information validity
2. Confusion of cause and effect
3. Complexity of effects, including nonlinearity, interaction, and dynamics

The questioning of information and the design of experiments is an integral part of SPC. Statistical techniques can be used to judge the fitness of data. Correlation analysis and cause and effect analysis can be used to quantify causal relationships. As complexity increases, there are many advanced methods to help with the solution of problems.

SPC techniques can be applied directly in the process industries along the interface where flowing product becomes discrete product, for example, where it is packaged. In the fluid process industries, where process dynamics play a big role, problem solution is confounded by a number of factors, such as the following:

- Variable interaction and inherent control loop cycling.
- Transport and mixing lags complicate correlation.
- Chemical and physical changes present many nonlinearities.
- Sampling rate independent of sample size or volume.
- Physical mixing and averaging of product in process.

Statistical techniques combined with control engineering principles can be applied to deal with these factors, as described in Chapter 1.

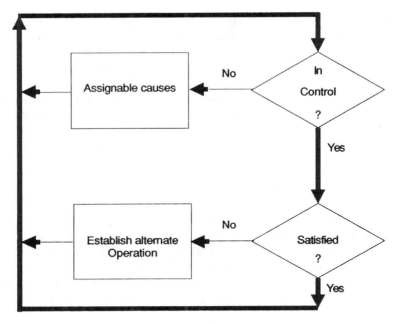

Figure 19-2   In Control and Satisfied, Continue Charting.

If the specification limits for the charted quality variable are outside its control limits, then one is in the very fortunate situation of being able to satisfy customers without any changes to the processing, production, or manufacturing system. Figure 19-2 depicts the process that must be followed. Charting of the variable must continue in order to ascertain that the process continues to be in statistical control and that the specifications are continuously met.

As time goes on, however, the need arises to tighten the specifications in order to meet competitive pressures and win market share. Therefore, SPC relates directly to cost and substantially to profits. Figure 19-3 shows this situation. The process is in statistical control but does not satisfy customer specifications. To do so requires further reduction in process variation. This is the *management loop*. Management needs to work on the total system to reduce this type of quality variation. The solution may require a variety of approaches, including process capability analysis, advanced statistical analysis, advanced control technique, and process design. An example of solving common-cause problems of this type will be presented later.

Figure 19-4 shows a variable that is not in statistical control because of a given cause or causes, which results in the violation of one or more of the rules. Rule violation is the result of assignable causes that are due to changes or variations of systematic variables, as discussed in Chapter 1. Examples of systematic variables are feed flows, feed compositions, temperatures, pressures, and resi-

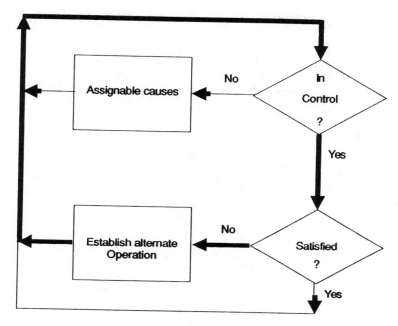

Figure 19-3 Problem Solving: The Management Loop.

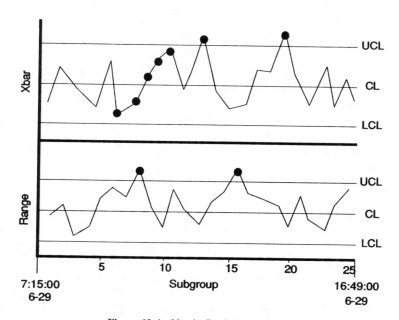

Figure 19-4 Not in Statistical Control.

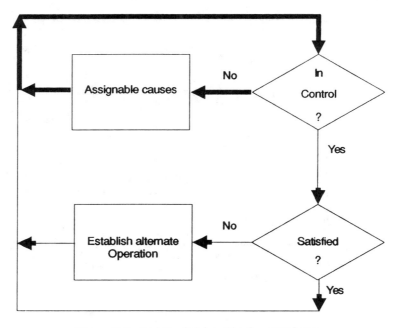

Figure 19-5   Problem Solving: The Operator Loop.

dence times in continuous processes, and faulty setup, defective raw material, and operator error in manufacturing processes.

Assignable causes are attributable to local manufacturing steps or local process units, and they can be typically eliminated by appropriate action by the local operators. This is the *operator loop* depicted in Fig. 19-5. The operator is responsible for eliminating or minimizing the assignable causes and for achieving statistical control. The remaining variation is then totally random, and any improvement is management's responsibility.

Problem Solving for Assignable Causes: The Operator Loop

A chemical plant example is used to illustrate how the tools of statistical process control can help solve problems ascribed to assignable causes. The chemical plant makes polymer. Figure 19-6 shows a *P* chart for the fraction of product rejected within a week, that is, the number of barrels of defective product divided by the number of barrels produced. This chart provides a picture of overall quality for a given week. Looking at the chart, one sees that there are times when the operator does very well (low fraction rejected), and there are other times that the fraction rejected is high and outside the control limits. What are the causes of rejection, and what action can eliminate the problem?

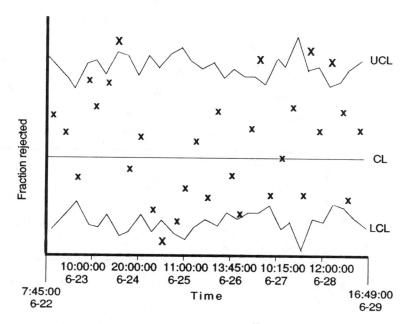

Figure 19-6   Total Production Fraction Rejected.

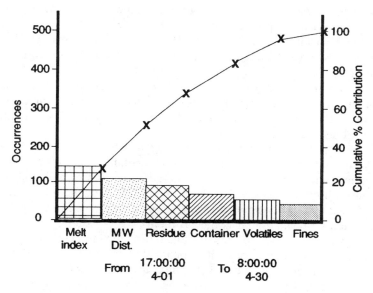

Figure 19-7   Prioritization of Rejection Causes for Polymer.

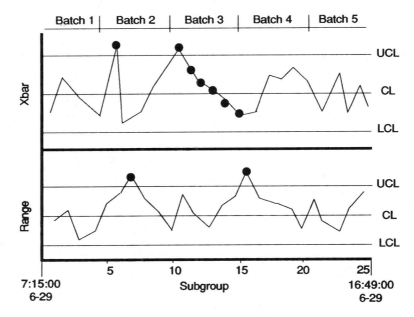

Figure 19-8   Melt Index xbar and Range Chart.

Figure 19-7 is a Pareto diagram that shows the causes of rejection for a given month. In this example, solving the melt index problem, results in 80% improvement in terms of product rejected. This suggests that one looks first at the melt index xbar and range chart, Fig. 19-8. The chart shows that the melt index is not in statistical control. Thus the melt index is a good candidate to pursue in solving the overall quality problem.

The operator, therefore, refers to the melt index cause and effect diagram of Fig. 19-9. This diagram shows the effect, the melt index, and all the possible causes that affect it, such as feed preparation variables, human factors, catalyst variables, and the reactors.

Typically, the percent of hydrogen, whose CUSUM chart is shown in Fig. 19-10, and the residence time of the reactants and catalyst, shown by the individuals chart in Fig. 19-11, are the usual causes of melt index problems. These charts, however, show that the percent of hydrogen and residence time variables are in statistical control and vary within acceptable limits, without exhibiting any unusual patterns. What is the problem?

Referring back to Fig. 19-8 and noting the intervals of catalyst batches, drawn in at the top of the chart, it appears that the melt index variation could be the result of differences in catalyst. The cause and effect diagram of Fig. 19-9 includes catalyst activity as a causal variable. The scatter diagram of Fig. 19-12 shows that the catalyst activity and melt index have a high positive correlation, rho = 0.932.

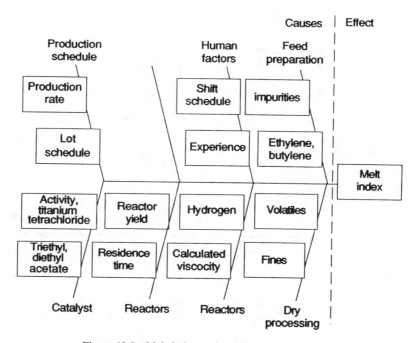

Figure 19-9   Melt Index and Its Causes of Variation.

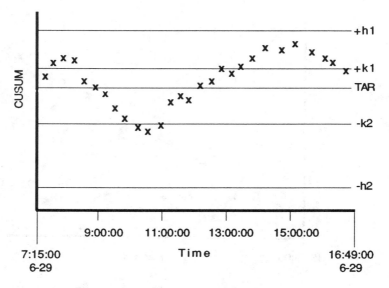

Figure 19-10   Percent Hydrogen CUSUM Chart.

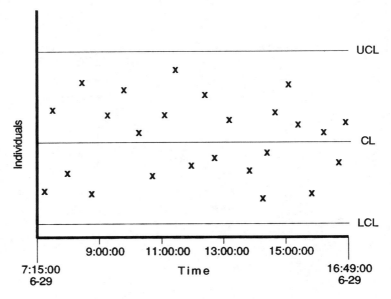

Figure 19-11   Residence Time Individuals Chart.

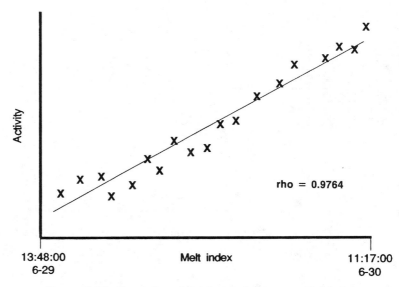

Figure 19-12   Correlation of Catalyst Activity versus Melt Index.

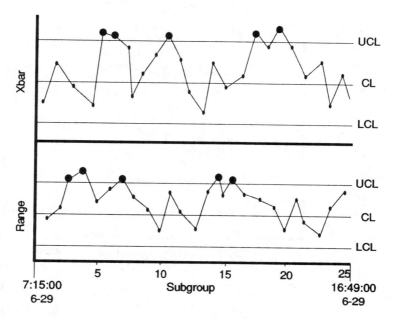

Figure 19-13   Catalyst Activity xbar and Range Chart.

Recall that rho = 1.0 is perfect correlation. The xbar and range chart for catalyst activity (Fig. 19-13) indicate significant batch-to-batch variation.

Suspecting problems in catalyst addition, the operator then prepares the plot of the titanium tetrachloride to monomer ratio shown in Fig. 19-14, which shows significant batch-to-batch effect. Therefore, the solution is to maintain this ratio more effectively. Subsequent investigations into catalyst preparation find two methods problems. The first is inaccurate weigh-out of ingredient; the other is inadequate mixing in the addition vessel.

This example shows how an operator can systematically use the charts and diagrams of statistical process control to identify and solve local problems due to assignable causes.

Problem Solving for Common Causes: The Management Loop

The process capability study is the primary discipline for dealing with common causes in SPC. Here, process refers to any combination of conditions that work together to produce a given result. Capability means the natural or undisturbed performance after extraneous influences are eliminated. In an industrial plant, the study may range from the analysis of a measuring device to a study of the quality and profitability of the entire plant.

An actual process example is used to demonstrate the application of the

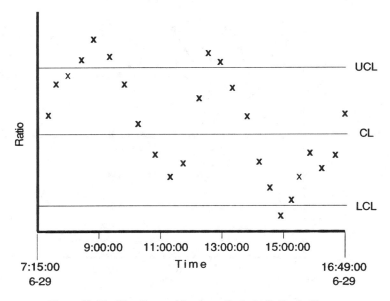

Figure 19-14   Titanium to Monomer Ratio Individuals Chart.

process capability analysis to the management problem of being in statistical control but not meeting product quality specifications. Reference 12 provides a description of the process, the process analysis, and the control strategies used to achieve the control system objectives of safe operation, stable control, increase in throughput, energy savings, and reduction in variation in product composition.

The process consists of approximately a dozen alcohol distillation columns. The methanol column is chosen to illustrate the approach. The feed to the column is overhead vapor from an upstream column. The feed composition is mostly methanol (MEOH) with a large amount of isopropyl alcohol (IPA) and water. The top product is methanol with about 30% IPA and as little water as possible. The bottom product contains the rest of the water and IPA in the feed, with as little methanol as possible.

Figure 19-15 shows a histogram for 28 samples taken once per day. The upper specification limit (*USL*) is 32% IPA, while the lower specification limit (*LSL*) is 24% IPA. The standard deviation (sigma) is 5.70%. The inherent process capability (CP) is

$$CP = \frac{USL - LSL}{6 \text{ sigma}} = \frac{8}{34.2} = 0.234 \qquad (19\text{-}2)$$

A process with a CP of less than 1.0 is not capable, because it cannot meet the required specification tolerance, whereas a process with a CP greater than 1.0 is indeed capable of meeting the required specification tolerance.

Therefore, the objective is to analyze the process phenomena and operations

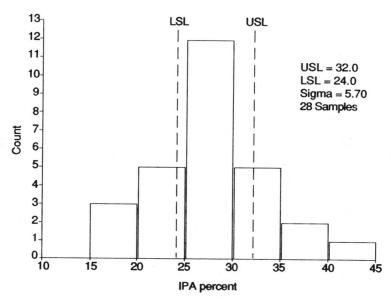

Figure 19-15   Methanol Column Top IPA Before.

methods to decide what changes will make the process capable. A detailed process study using basic principles of chemistry and chemical engineering reveals that one major cause of the problem is that the process reverses. The cause and effect variables change from being positively to being negatively correlated depending on the amount of water in the feed.

Typically, increasing the heat to the column increases the volatility and mass flow of methanol to the top of the column. However, in this process, when the water content of the feed increases due to problems in the upstream column, the relative volatility of IPA becomes larger; further increases in heat drive IPA up the column, while increasing methanol concentration in the bottom product. This is the reverse of what is expected. The solution to the problem is to use chromatographic analyzers at the appropriate streams together with a control strategy to keep the process stable, and, when the instability occurs, to return it to the stable side as fast as possible.

The study also indicates that the existing control system needs improvement in order to reduce the variation to acceptable levels. Therefore, the decision is to replace the existing pneumatic controls with a distributed control system and to install chromatographic analyzers.

Figure 19-16 shows a plot of top IPA concentration for the 28 daily samples before and 72 hourly average samples after the installation and implementation of the new control system. The 72 hours of data contain set-point changes, particularly in the first 24 hours. The advanced controls force the column to steady

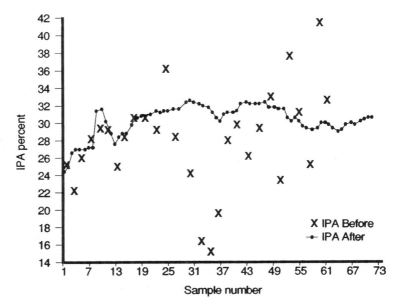

Figure 19-16   Trend of Methanol Column Top IPA After.

state in a much shorter time than 72 hours. Thus a longer time period will probably show a similar response. Also, since the process takes more than 1 hour to return the measurement to set point, the comparison of daily spot samples to hourly averages seems acceptable.

The histogram of Fig. 19-17 shows the last 48 hourly samples, thus eliminating the first 24 samples that contain the set-point changes. The upper and lower specifications are the same as before. The standard deviation is 1.07. Therefore, the process capability is

$$CP = \frac{8}{6.42} = 1.246 \tag{19-3}$$

Therefore, the new control system makes the process capable of achieving its specification tolerance. Moreover, the reduction in variation from a standard deviation of 5.7% to 1.07% allows an increase in the average IPA concentration in the top product. Together with the stabilization of the column, this results in a 16% increase in throughput and more than 10% energy savings, in addition to the reduction in the variation of IPA concentration in the top product.

The preceding example illustrates the fact that reduction in variation due to common causes requires management's attention and substantial investment in new sensors and control system in order to achieve the required objectives. Although the new control system and sensors are sufficient to solve the problem, as in this case, many times the process itself may need to be redesigned or replaced

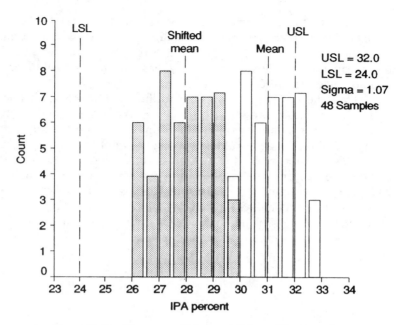

Figure 19-17   Histogram of Methanol Column Top IPA After.

with a new process. Moreover, operating methods and procedures throughout the production system may require modification in order to achieve acceptable common-cause variation.

This requires more in-depth process understanding and analysis. This approach, termed *phenomenological,* is characterized by the use of physics, chemistry, mathematics, engineering, and correlated variables. The statistical analysis approach, on the other hand, treats the process as a black box with cause and effect variables and is characterized by the use of data, mathematics, statistics, and random variables. The two approaches work together. The statistical approach indicates that problems exist as a result of assignable causes, while the phenomenological approach typically is needed to solve the problems.

19.5 SUMMARY

Statistical process control enables the identification of problems and helps in their resolution. SPC first determines if the variables and the process are in statistical control. If not, the operator is responsible. The operator uses the statistical tools of charts and diagrams to navigate through the possible causes of the problem and takes the required actions to solve it.

If the variables and the process are in statistical control, but performance is not satisfactory, alternate operation needs to be established. This is the responsibility of management. The quality improvements for common causes require not only statistical tools but also process understanding, which includes basic principles of physics, chemistry, engineering, control theory, mathematics, and advanced statistics.

Thus the statistical approach and the phenomenological approach are used in a complementary way to identify and solve problems. The integration of statistical process control tools with traditional control tools provides a much stronger foundation for problem solving than either alone.

This chapter solves Eq. (19-1) for quality improvement. Participative management sets clear goals for the future and is actively involved in attaining them. The tools of SPC consist of charts for variables, charts for attributes, and diagrams for cause and effect analysis. And, finally, SPC tools, along with traditional control tools, help solve problems. The implementation of statistical quality control results in improving quality, which is indisputably followed by increased productivity.

Chapter 20

Approaches to Implementation in the Oil and Gas Industry

The purpose of this chapter is to use the refinery as an example of how to approach the implementation of statistical process control in the oil and gas industry. The three major objectives of this chapter are as follows:

1. To provide a process description for the refinery.
2. To provide the cause and effect diagram and typical cause and effect variables to be charted for each major unit of the refinery.
3. To provide the refinery cause and effect hierarchy.

20.1 REFINERY PROCESS DESCRIPTION

Reference 13 is the main reference for this and subsequent sections. Figure 20-1 shows a typical refinery diagram with the major units and products. The crude oil is heated in a furnace, and from there it enters the atmospheric crude distillation tower, which separates the crude under atmospheric pressure into a number of product streams. The top product contains butanes and lighter wet gas and goes to the gas plant. The straight run gasoline product goes to treating and blending, while heavy naphtha becomes feed to the reforming unit. The kerosene product

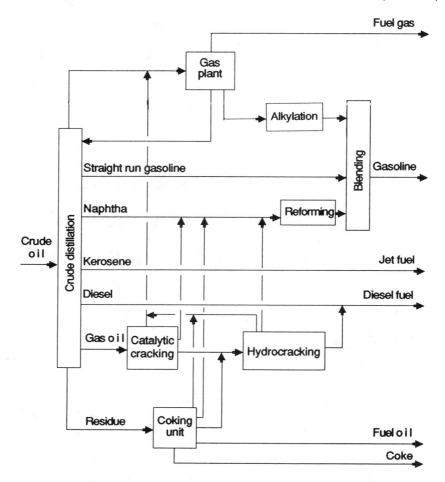

Figure 20-1   Refinery Diagram.

is used for jet fuel and the diesel is used for diesel oil. The heavy gas oil becomes feed to the catalytic cracking unit. The bottom product of the atmospheric tower is the topped crude, which is fed to the vacuum crude distillation tower, which separates it into a top vacuum gas oil product and a bottom residue.

The bottom residue is the feed to the coking unit, which thermally cracks it into wet gas, coker gasoline, light coker gas oil, heavy coker gas oil, and coke. The light coker gas oil is used as fuel oil.

The crude distillation gas oil is the feedstock to the catalytic cracking unit, which cracks it into light components, such as ethane, propane, and butane; the major products are catalytic gasoline and cycle oil. The cycle oil from catalytic cracking and the heavy coker gas oil are feedstock to hydrocracking, which cracks it into light ends, gasoline, and middle distillates used as petrochemical feed.

Light ends from crude distillation and the coking and cracking units are fed to the gas plant. The gas plant fractionates its feeds into fuel gas, liquefied petroleum gas (LPG), normal butane, isobutane, propylene, butelene, and other unsaturated hydrocarbons. The fuel gas is burned in refinery furnaces, normal butane is used in gasoline blending, and the unsaturated hydrocarbons (olefins) and isobutane are fed to the alkylation unit. The alkylation unit converts olefins and isobutane into the high-octane alkylate product, which is blended into premium gasoline.

The reforming unit takes the gasoline (naphtha) streams from crude distillation and the coking and cracking units and converts them into reformate product, which is a higher-octane product and is blended into regular and premium gasolines.

20.2 CAUSE AND EFFECT DIAGRAM FOR CRUDE DISTILLATION

This section provides a short description of the crude distillation unit, some typical control strategies, and some cause and effect variables for it.

Crude oil is composed of a great number of components, ranging from dissolved gases, such as hydrogen and methane, to very heavy molecules with a large number of carbon atoms in their molecular structure. In addition, it contains sulfur, salt, and other impurities.

Crude distillation consists of the atmospheric and vacuum towers. Crude oil is heated by heat recovered from downstream units and the furnace to a desired temperature and introduced to the atmospheric tower, which separates it into different product streams under atmospheric conditions. The bottom topped crude stream is fed to the vacuum tower for further separation under lower pressure, and hence temperature, than that of the atmospheric tower in order to avoid coke deposits on the towers. The products of crude distillation are shown in Fig. 20-1.

Crude oils and the heavy sidestream products of crude distillation are characterized by their temperature versus percent distilled by volume, with a temperature range of 80° to 1000° Fahrenheit (F) for a typical crude. This is referred to as the boiling point curve. For each crude distillation product, a boiling point curve is usually obtained every 8-hour shift. The initial boiling point (IBP) and the end point (EP), which is the upper temperature limit of the boiling point curve, are two temperatures of particular interest.

Reference 14 discusses typical control strategies for the atmospheric tower, heat removal control, and distillation control in general. Usually, the temperature at which 5% of a product is vaporized is the IBP, and the temperature at which 95% is vaporized is the EP. Product specification is usually the end point.

The temperature of the crude oil is maintained at a desired value, approximately 700°F, to have 50% to 70% of the crude charge vaporized as it enters the atmospheric tower. The fuel flow to the heating furnace is controlled to maintain the heat flow to the furnace at a desired value. The required heat flow is set based on crude oil flow and its temperature at the furnace inlet and outlet. An outlet temperature controller maintains the furnace outlet temperature at the desired set point by manipulating the outlet temperature used in the heat flow computation.

Each crude distillation sidedraw product is withdrawn under flow control. The flow measurement, which is the controlled variable of the flow controller, is the sum of the sidedraw flow of interest plus the flow of all the lighter products. The flow controller set point is set in ratio to crude flow. The ratio is adjusted by the end-point analyzer. Moreover, the steam flow to the sidestream stripper is set in ratio to its sidedraw product flow.

Other controls consist of material and energy balances that calculate the internal liquid and vapor flows to maintain the internal traffic just below the flooding limit. Percent of flooding is controlled by adjusting the set point of the upper pumparound heat flow removal controller. Similar controls are used for the vacuum tower heater and distillation tower.

The cause and effect diagram for the crude distillation unit can be built by charting the following cause and effect variables.

EFFECT VARIABLES

- Straight run gasoline flow and end point
- Naphtha flow and end point
- Kerosene flow and end point
- Diesel flow and end point
- Gas oil flow and end point
- Other lab analysis variables, for example, pour and cloud point and viscosity
- Other measured variables
- Crude distillation performance calculation, such as energy used per barrel or pound of crude charged, value added per unit of crude charged

CAUSE VARIABLES

- Crude oil flow, boiling point curve, and temperature inlet to crude tower
- Reflux flow
- Calculated percent of flooding
- Top and bottom pumparound heat flows
- Overhead vapor temperature
- Ratio of gasoline plus lighter material to crude flow
- Ratio of naphtha plus lighter material to crude flow
- Ratio of kerosene plus lighter material to crude flow
- Ratio of diesel plus lighter material to crude flow

- Ratio of gas oil plus lighter material to crude flow
- Topped crude temperature inlet to vacuum tower
- Other atmospheric tower temperatures
- Vacuum tower pressure and temperatures
- Other measured and analysis variables

The cause and effect variables are charted using the procedures presented in Chapter 17 and are attached to a cause and effect diagram, such as the one shown in Fig. 15-1. Real-time SPC monitoring and plant operation then follows, as described in Chapter 18.

20.3 CAUSE AND EFFECT DIAGRAM FOR CATALYTIC CRACKING

Gas oil from the atmospheric and vacuum crude towers and the coking unit is introduced as feed to the catalytic cracking unit. The unit consists of the reactor/regenerator and the fractionator.

Heated feed and regenerated catalyst enter the cylinder, where the reaction starts, and proceed to the reactor, where the reaction is completed. As the cracking reaction proceeds, catalyst activity is continuously reduced by coke formation on its surface. The hydrocarbon vapors and catalyst are separated, and the hydrocarbon vapors are taken from the top of the reactor to the fractionator for separation. After steam stripping to remove any oil left on it, the spent catalyst is returned to the regenerator.

The catalyst is regenerated by burning the coke deposits with air and is then returned to contact fresh feed again. The products of combustion are removed by the regenerator flue gas.

The fractionator separates the oil vapors into catalytic gasoline, light cycle oil, heavy cycle oil, decanted oil, and heavy slurry bottoms.

Feed temperature is controlled using a heat rate controller similar to that for the crude tower furnace heater, with fuel flow manipulated to control the temperature at the furnace outlet to a desired set point. A temperature difference controller across the regenerator bed of spent catalyst controls air flow to the regenerator to maintain the temperature difference at a desired set point. Differential pressure between the reactor and regenerator is controlled to a desired value by manipulating regenerator flue gas flow. Catalyst flow to the regenerator is controlled to maintain the level of the catalyst bed in the reactor at a desired set point.

The reactor temperature is maintained to a desired set point by manipulating catalyst flow to the reactor. A conversion controller uses the calculated percent of conversion and adjusts the set point of the reactor temperature controller to maintain conversion at a desired value.

Mass and energy balance models have been developed for some fluid catalytic cracking reactors and regenerators. They are used to monitor their performance and also for predicting what happens when the flow and composition of the reactants and reaction model parameters vary. When used to monitor cracking and catalyst regeneration reactions during unit operation, the difference between model-predicted variables and their actual values is charted to monitor model performance.

The gasoline product of the fractionator is controlled by manipulating reflux flow. A top temperature controller sets the reflux flow set point, while an end-point controller adjusts the temperature set point to maintain the gasoline end point at a desired value. The accumulator level sets gasoline product flow.

The light cycle oil end point is controlled using an end-point controller that manipulates flow to the stripper. Stripper steam flow is controlled in ratio to oil flow. The stripper level sets the light cycle oil product flow.

The cause and effect diagram for the catalytic cracking unit can be built by charting the following cause and effect variables.

EFFECT VARIABLES

- Catalytic gasoline flow and end point
- Light cycle oil flow and end point
- Heavy cycle oil flow and end point
- Decanted oil flow
- Bottom slurry flow
- Reactor and regenerator performance
- Other measured and analysis variables
- Model-predicted variables
- Catalytic cracking unit overall performance and value added

CAUSE VARIABLES

- Feed flow, temperature, and end point
- Furnace heat flow
- Reactor temperature and conversion
- Regenerator temperatures and temperature difference
- Catalyst circulation rate
- Catalyst to feed ratio
- Flue gas oxygen, carbon dioxide, and carbon monoxide
- Fractionator pressure and temperatures
- Other measured and analysis variables

20.4 CAUSE AND EFFECT DIAGRAM FOR COKING

The residue from the bottom of the vacuum tower is heated by a furnace and fed into the coke drums, where it is thermally cracked into gas, coker gasoline, light coker gas oil, heavy coker gas oil, and coke. The vapor from the coke drums is separated into the product streams by the fractionator. The fractionator bottoms are recycled to the coke drums.

Temperature, pressure, and recycle ratio, that is, furnace flow to fresh feed flow ratio, are the three main operating variables that determine product quality and yield. The temperature is controlled by adjusting the heat flow controller set point. The heat flow controller in turn sets fuel flow. Coke drum pressure is controlled by setting cooling water flow. The recycle ratio is controlled by adjusting furnace flow. Coker gasoline end point control is achieved using the fractionator overhead temperature to set reflux flow. Gasoline product flow is set by the accumulator level.

The light coker gas oil end point is controlled using a vapor temperature controller at a tray halfway between the light and heavy oil draw trays to set light oil flow. The heavy coker gas oil end point, on the other hand, is controlled using a vapor temperature at the pumparound tray to set pumparound return flow.

The cause and effect diagram for the coking unit can be built by charting the following cause and effect variables.

EFFECT VARIABLES

- Coker gasoline flow and end point
- Light coker gas oil flow and end point
- Heavy coker gas oil flow and end point
- Decanted oil flow
- Coke drum performance
- Other measured and analysis variables
- Coking unit overall performance and value added

CAUSE VARIABLES

- Feed flow and temperature
- Furnace heat flow
- Coke drum pressure
- Recycle ratio
- Reflux flow
- Pumparound flow
- Fractionator pressure and temperatures
- Other measured and analysis variables

20.5 CAUSE AND EFFECT DIAGRAM FOR HYDROCRACKING

The catalytic cracking unit uses the more easily cracked part of the gas oil from the crude distillation unit. The catalytic hydrocracker receives the heavy cycle oil from the catalytic cracking unit and the heavy coker gas oil and hydrocracks them using higher pressures and a hydrogen atmosphere in the presence of catalyst.

The hydrocracking reaction is exothermic, and the catalyst bed inlet temperature is controlled by adjusting the hydrogen quench flow. The catalyst bed outlet temperature sets the catalyst bed inlet temperature set point to maintain outlet temperature at a desired value.

After cooling, the reactor effluent goes to a high-pressure separator, which separates the hydrogen-rich gases from liquid product. The gases are recycled and mixed with hydrogen makeup in the compressor. Hydrogen makeup flow is manipulated to maintain compressor outlet pressure, and hence reactor pressure, at a desired value.

The compressed gases are then mixed with fresh feed as they enter the heater. Heater outlet temperature is controlled by setting fuel flow. Then the heated feed mixed with hydrogen-rich gas enters the reactor. Liquid from the reactor goes to the fractionator, where it is separated into an overhead light ends, straight run gasoline for blending, naphtha for reforming, and diesel fuel. Fractionator control is similar to that for the catalytic cracking unit fractionator.

The cause and effect diagram for the hydrocracking unit can be built by charting the following cause and effect variables.

EFFECT VARIABLES

- Straight run gasoline flow and end point
- Naphtha flow and end point
- Diesel flow and end point
- Other measured and analysis variables
- Hydrocracking unit overall performance and value added

CAUSE VARIABLES

- Feed flow and temperature
- Furnace heat flow
- Reactor pressure and inlet and outlet temperatures
- Reflux flow
- Fractionator pressure and temperatures
- Other measured and analysis variables

20.6 CAUSE AND EFFECT DIAGRAM FOR REFORMING

Crude tower naphtha, coker gasoline, and gasoline from the cracking units are fed to the catalytic reformer to improve their octane numbers. Usually, there are three reactors in series and feed is heated before it enters each reactor.

The first reactor outlet temperature is controlled to a desired value by setting the fuel to feed flow ratio, while the outlet temperature of the other two reactors is controlled by directly manipulating fuel flow. In addition, the first reactor steam flow is set in ratio to feed flow or carbon flow when in-line analyzers are used.

The last reactor effluent after cooling goes to a high-pressure separator, which separates the hydrogen-rich gases from liquid product. The gases are split into a hydrogen recycle stream and a hydrogen by-product stream. Hydrogen recycle flow is manipulated to maintain hydrogen recycle compressor outlet pressure, and hence reactor pressure, at a desired value.

The compressed gases are then mixed with fresh feed as they enter the heater. The heater outlet temperature is controlled by setting the fuel flow. Then the heated feed mixed with the hydrogen-rich gas enters the reactor. Liquid from the reactor goes to the fractionator, where it is separated into an overhead product, which is mainly butanes and the reformate bottoms product.

Typical control consists of setting the reboiler steam flow to maintain a heat flow to feed flow ratio. The ratio is adjusted by the reboiler outlet temperature controller to maintain the temperature at a desired value.

Combined reflux and distillate control, described in reference 15, is used to control the top of the depropanizer. The top pressure controller sets the combined reflux plus distillate flow. The ratio of distillate to distillate plus reflux flow is set by the operator or, when available, by an in-line analyzer.

The cause and effect diagram for the reforming unit can be built by charting the following cause and effect variables.

EFFECT VARIABLES

- Reformate flow and octane number
- Butane flow
- Other measured and analysis variables
- Model-predicted variables
- Reforming unit overall performance and value added

CAUSE VARIABLES

- Feed flow and temperature
- Furnace heat flow
- Steam flow
- Reactor pressure and outlet temperatures
- Reflux flow

- Fractionator pressure and temperatures
- Other measured and analysis variables
- Model-predicted variables

20.7 CAUSE AND EFFECT DIAGRAM FOR ALKYLATION

Refinery alkylation processes use either hydrogen fluoride or sulfuric acid as catalyst to react mainly butylenes with isobutane into an isooctane alkylate product, which is a high-octane gasoline. Butylenes and other similar molecules with at least two carbon atoms joined by double bonds are referred to as olefins. Reference 16 describes the control strategies used in alkylation units.

Olefin feed contains propane, normal butane, and isobutane. The bulk of isobutane is recycled from the deisobutanizer, and there is also a relatively small isobutane makeup stream. The total isobutane is calculated, and olefin feed flow is controlled to maintain a desired isobutane to olefin ratio.

Reactor pressure is controlled at a set point high enough to maintain the reactants at liquid phase. In sulfuric acid units, the average reaction temperature is controlled by setting the refrigeration compressor speed.

The depropanizer separates propane from isobutane. Typical control consists of setting reboiler steam flow to maintain a heat flow to feed flow ratio. The ratio is adjusted by an analysis controller to maintain the percent of propane impurity at the depropanizer bottoms stream at a desired value.

Combined reflux and distillate control, described in reference 15, is used to control the top of the depropanizer. The top pressure controller sets the combined reflux plus distillate flow. The percent of isobutane analysis controller in the top vapor stream sets the ratio of distillate to distillate plus reflux flow.

The deisobutanizer separates isobutane as top product, normal butane as a sidestream product, and the bottom alkylate product. A top pressure controller sets isobutane distillate flow, with reflux flow set in ratio to distillate flow. An analysis controller for percent of isobutane impurity in the normal butane stream sets the normal butane flow. The alkylate product is withdrawn under bottoms level control, and the reboiler outlet temperature sets reboiler steam flow.

The cause and effect diagram for the alkylation unit can be built by charting the following cause and effect variables.

EFFECT VARIABLES

- Alkylate flow and octane number
- Normal butane flow and percent of isobutane in normal butane
- Propane flow and percent of isobutane in propane
- Other measured and analysis variables

- Model-predicted variables
- Alkylation unit overall performance and value added

CAUSE VARIABLES

- Reactor pressure and temperature
- Olefin to isobutane ratio
- Total isobutane flow
- Total olefin flow
- Depropanizer pressure, top temperature, and reflux flow
- Depropanizer bottoms percent of propane
- Deisobutanizer pressure and reboiler temperature
- Other measured and analysis variables
- Model-predicted variables

20.8 CAUSE AND EFFECT DIAGRAM FOR GAS PLANT

Gas from crude distillation, the coking unit, and the cracking units is fed to the gas plant for the recovery of propane and heavier components and for the production of desulfurized fuel gas consisting mostly of methane and ethane.

According to reference 13, the gas is first compressed and fed into an absorber–deethanizer. In the top absorption section of this column, lean absorption oil absorbs most of the propane and all heavier components and brings them to the bottom stripping section. Vapor from the top of the absorber is fed into the sponge absorber, where lean sponge oil is used to absorb any of the vaporized hydrocarbons. The rich sponge oil is then returned to the column that it came from as lean sponge oil.

The bottom striping section removes ethane and methane from the bottoms liquid product. This liquid product then is fed to the debutanizer column. The debutanizer separates propane and propylene and butanes and butylenes as top product and pentanes and heavier components as naphtha bottoms product. The naphtha is fed to the naphtha splitter and separated into an overhead light, straight run gasoline, which is desulfurized and used in gasoline blending. Most of the bottoms lean absorption oil product is recycled to the absorber section of the absorber–deethanizer column, and the excess is fed to a hydrotreater and reformer. The overhead product from the debutanizer is desulfurized and fed to the depropanizer, where it is separated into propane overhead product and butanes bottom product.

References 14 and 15 provide details of the control strategies used in the units that comprise the gas plant. Control of the absorber section of the absorber–deethanizer column consists of controlling lean oil flow in ratio to gas flow, with the ratio being set by the operator. The stripper section is controlled to a desired temperature two or three trays above the bottom of the column by setting reboiler heat flow, which in turn sets steam flow.

Debutanizer top control uses pressure to set the reflux plus distillate flow. Distillate flow is set by the ratio of distillate to reflux plus distillate, and the ratio is adjusted by the top vapor temperature controller to maintain temperature at a desired value. Reboiler outlet temperature controls the flow of steam to maintain temperature at a desired value. The naphtha splitter control is similar to that of the debutanizer.

The depropanizer top control is also similar to that of the debutanizer top, but, in addition, an isobutane analysis controller sets the temperature controller set point. Morever, the temperature measurement is made a few trays below the top of the column. A temperature two or three trays above the bottom of the column sets heat flow.

The cause and effect diagram for the gas plant can be built by charting the following cause and effect variables.

EFFECT VARIABLES

- Gas flow and composition
- Propane flow and composition
- Butane flow and composition
- Light straight run gasoline flow and octane number
- Other measured and analysis variables
- Gas plant overall performance and value added

CAUSE VARIABLES

- For each column, pressure and temperatures
- For each column, feed flow and feed composition
- Other measured and analysis variables

20.9 CAUSE AND EFFECT DIAGRAM FOR BLENDING

According to reference 13, blending is currently done in line with computers controlling the blending of the various base stocks and additives. The computer database maintains blending stock inventories, cost, and physical property data and uses them with linear programming optimization models to blend products to the required specifications at lowest cost. The base stocks are the products of the refinery units discussed previously. Typical desired blending specifications are Reid vapor pressure (RVP), octane number, and pour point.

The cause and effect diagram for blending can be built by charting the following cause and effect variables.

EFFECT VARIABLES

- For each product, flow, specifications, and cost

CAUSE VARIABLES

- For each base stock, flow and composition
- For each additive, flow and properties
- Linear programming or other blending model parameters
- Other measured and analysis variables

20.10 REFINERY CAUSE AND EFFECT HIERARCHY

A cause and effect diagram such as the one shown in Fig. 15-1 is built for each refinery unit discussed in the preceding sections. The types of charts to be used for the cause and effect variables are selected based on the analysis procedures presented in Chapter 17.

The unit cause and effect diagrams are organized into the refinery cause and effect hierarchy shown in Fig. 20-2, based on the discussion in Chapter 15. As pointed out in that chapter, when necessary the units can be divided into subunits with their own cause and effect diagrams, or a number of units and subunits can be organized into area cause and effect hierarchies.

The refinery cause and effect hierarchy is then used to monitor and operate the whole refinery with the aim of continuously providing products on specification at minimum cost. It provides all information on cause and effect and aids in the solution of problems.

The same hierarchy can be used by everybody in the refinery, from the refinery superintendent to the operators, from the instrumentation and engineering managers to the instrument technicians and the unit engineers, and from information system management to the programmers.

Moving up the hierarchy, information is abstracted to provide what is necessary at that level, no more and no less. Take, for example, the refinery cause and effect diagram of Fig. 20-2, which is the top-level diagram in the hierarchy. For this diagram, the charts and other analysis diagrams and text objects attached to the effect box provide monitoring and all information for all product yields and specifications and for the overall performance of the refinery. The cause boxes provide access to each major unit's cause and effect diagram to zoom in for more detail when necessary. However, the empty cause boxes can be used to attach charts and other objects that furnish abstracted information about the most important cause and effect variables for each unit.

Thus, one and all can navigate electronically through all available information from feeds to products in the areas of their responsibility. Moreover, they can also access and navigate through areas that affect them and areas they affect

Refinery Cause and Effect Diagram

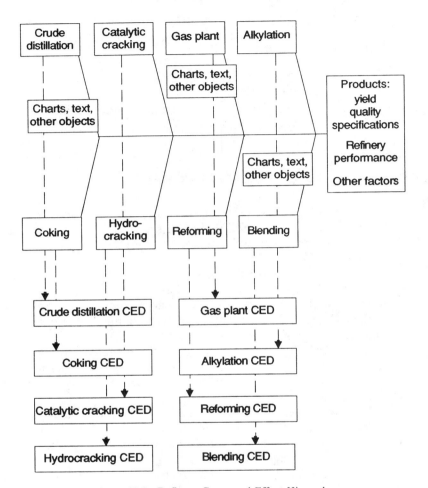

Figure 20-2   Refinery Cause and Effect Hierarchy.

in order to monitor operation and performance continuously and, therefore, strive for continuous improvement and enhanced productivity.

Finally, the refinery cause and effect hierarchy makes everybody continuously sensitive to internal as well as external customers and suppliers. This results in a continuous commitment to maintain and improve customer satisfaction, to improve operation, and thus to improve productivity.

Chapter 21

Approaches to Implementation in the Chemical Industry

The purpose of this chapter is to use the ethylene plant as an example of how to approach the implementation of statistical process control in the chemical industry.

The three major objectives of this chapter are as follows:

1. To provide a process description for the ethylene plant.
2. To provide the cause and effect diagram and typical cause and effect variables to be charted for each major unit of the ethylene plant.
3. To provide the ethylene plant cause and effect hierarchy.

21.1 ETHYLENE PROCESS DESCRIPTION

There are many different processes for producing ethylene and individual plants differ. Reference 17 provides flowsheets for some ethylene plants, one of which is used in the discussion here. Figure 21-1 shows a typical ethylene plant diagram with the major units and products.

Cracking furnaces convert ethane, propane, naphtha, or gas oil feeds to products, primarily ethylene, along with some propylene, butadiene, and heavier

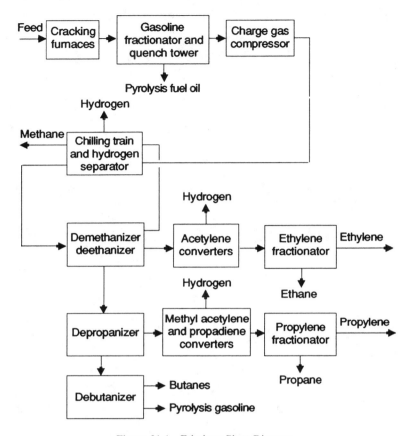

Figure 21-1   Ethylene Plant Diagram.

hydrocarbons. The furnace effluent goes to the fractionator where pyrolysis fuel oil is taken out as bottoms product. It is then cooled in the quench tower, compressed, and sent to the recovery section.

Unreacted methane and ethane recovered by the distillation columns are recycled back to the cracking furnaces. Refrigeration systems using ethylene and propylene as refrigerants provide cooling and reboil heat for the distillation columns. Acetylene, methyl acetylene, and propadiene impurities are hydrogenated in the converters and thus removed from the product streams.

The ethylene and propylene products are separated in the ethylene and propylene fractionators, respectively. The heavier hydrocarbons are separated in the debutanizer into an overhead butanes product and the bottoms pyrolysis gasoline.

21.2 CAUSE AND EFFECT DIAGRAM FOR CRACKING FURNACES

Typically, there are a number of cracking furnaces in an ethylene plant. This section provides a short description of a cracking furnace, some typical control strategies, and some cause and effect variables for the process.

The cracking furnace has a number of parallel cracking coils (tubes). Hydrocarbon and steam are mixed and preheated in the convection section at the top of the furnace. Then the mixture enters the radiant section, where fuel burners heat it further and the hydrocarbons are cracked into products. Coil outlet temperature is controlled to a desired value by manipulating the feed flow to the furnace.

Conversion is defined as the fraction of ethane or propane converted to products, and it is computed based on chromatographic analysis of the furnace feed and effluent streams. A conversion controller sets the coil outlet temperature controller set point to maintain conversion at a desired value.

Since the feed is used to control coil outlet temperature, the furnace throughput is determined by setting the heat flow to the furnace. Heat flow is computed using chromatographic analysis or the density of the fuel and its flow. The throughput controller uses as its measurement the total flow to the furnace, and its set point is the desired throughput flow. The throughput controller sets the set point of the heat flow controller, which in turn sets fuel flow.

When the fuel is gas and the gas flow to each furnace zone is controlled by burner pressure, then the heat flow controller sets the set points of the burner pressure controllers. The average temperature of all heating zones of the furnace is balanced by providing bias adjustments for each fuel flow or burner pressure controller.

Dilution steam is added to the feed to increase the yield of olefins by lowering the hydrocarbon partial pressure. This partial pressure reduction also lessens the tendency to form coke in the cracking coils and to deposit tar in the downstream heat exchange surfaces.

Steam flow is set in ratio to feed flow. The higher the steam to feed ratio for a given feed flow, the faster the mixture goes through the tubes. Therefore, the residence time decreases and the conversion decreases. The opposite is true when the steam to feed ratio is lowered.

As time goes on, coke deposits in the coils and their cracking efficiency decreases. Therefore, to maintain conversion, the coil outlet temperature has to be increased or the steam to feed ratio has to be reduced within limits. After a period of continuous operation, say, one to two months, the furnace is taken off line to remove coke deposits in the coils and heat exchange surfaces.

The cause and effect diagram for the cracking furnaces can be built by charting the following cause and effect variables.

EFFECT VARIABLES FOR EACH FURNACE

- Furnace effluent composition
- Conversion
- Steam generation
- Other lab analysis variables
- Other measured variables
- Cracking furnace performance calculation, such as energy used per unit feed and value added per unit of feed charged

CAUSE VARIABLES FOR EACH FURNACE

- Feed flow for each pass and composition
- Steam flow for each pass and steam to feed ratio
- Coil outlet temperatures
- Pressure for each zone
- Furnace heat flow
- Other measured and analysis variables

The cause and effect variables are charted using the procedures presented in Chapter 17 and are attached to a cause and effect diagram, such as the one shown in Fig. 15-1. Real-time SPC monitoring and plant operation then follow as described in Chapter 18.

21.3 CAUSE AND EFFECT DIAGRAM FOR COMPRESSORS

The raw gas from the quench tower is compressed and then dried and sent to the demethanizer. The gas compressor and the ethylene and propylene refrigeration system compressors are usually driven by steam turbines. The steam turbine speed is controlled by manipulating steam to the compressor to maintain pressure in the downstream units. The compressors are protected from surging by antisurge control systems. Antisurge control uses the suction orifice differential and the differential pressure across the compressor and manipulates the bypass flow from discharge to suction to keep the compressor on the safe side of surge.

The objective is to operate the plant so that the compressor recycle flows are minimized, which in turn minimizes energy use. On the other hand, this has to be balanced by the economics of the rest of the plant, because there is interaction between pressure and the operation of the cracking furnaces and the fractionation columns.

The cause and effect diagram for the compressors can be built by charting the following cause and effect variables.

EFFECT VARIABLES

- Gas compressor discharge pressure
- Ethylene compressor discharge pressure and temperature
- Propylene compressor discharge pressure and temperature
- Other measured and analysis variables
- Plant economic model-predicted variables
- Overall performance of compressors and refrigeration

CAUSE VARIABLES

- Gas compressor pressures and flow differential
- Ethylene compressor pressures and flow differential
- Propylene compressor pressures and flow differential
- Other measured and analysis variables
- Other measured, model-predicted, and analysis variables

21.4 CAUSE AND EFFECT DIAGRAM FOR CONVERTERS

Typically, there are two acetylene converters that hydrogenate acetylene and thus remove it from the ethylene product. This is achieved with the appropriate conditions of temperature and hydrogen and carbon monoxide addition.

The inlet temperature to each reactor is controlled by manipulating the reactor feed flow or steam flow and/or cooling water to the heat exchangers. The product from each reactor is analyzed, and the analysis is used to control the acetylene in the effluent streams by setting the set point of the inlet temperature controllers.

Hydrogen is added in ratio to acetylene in the feed by analyzing the feed and hydrogen streams and manipulating hydrogen flow to maintain the ratio at a desired value. The methyl acetylene and propadiene converters hydrogenate methyl acetylene and propadiene to remove them from the propylene product. The control strategy for these converters is similar to that for acetylene converters.

The cause and effect diagram for the converters can be built by charting the following cause and effect variables.

EFFECT VARIABLES

- Acetylene in product and conversion
- Methyl acetylene and propadiene in product and conversion
- Other measured, calculated, and analysis variables
- Overall performance of acetylene converters

CAUSE VARIABLES

- For each reactor feed flow, composition and temperature
- Hydrogen flow and hydrogen to acetylene calculated ratio
- Hydrogen to methyl acetylene plus propadiene calculated ratio
- Other measured, model-predicted, and analysis variables

21.5 CAUSE AND EFFECT DIAGRAM FOR OTHER COLUMNS

This section covers all fractionation columns except the ethylene and propylene fractionators, which are the main product columns and are organized with their own cause and effect diagrams for emphasis. References 14 and 15 provide details of the control strategies used for the fractionation columns here.

The furnace effluent enters the fractionator, which separates it into a gasoline and lighter top product and the pyrolysis oil bottoms product. The top temperature of this column is controlled by manipulating reflux flow. A gasoline end-point analyzer adjusts the top temperature controller set point to maintain the gasoline end point at a desired value. The bottoms product is set by the bottoms level.

The demethanizer column removes as top product methane and lighter components, which are recycled back to the furnace. Column pressure sets the methane vapor product. The concentration of ethylene in the top product is held more constant by setting the reflux flow in ratio to the methane vapor flow. If an analyzer is used for control of the ethylene concentration in the methane product, then the ethylene analysis controller sets the ratio.

A temperature five to ten trays from the bottom sets heat flow. The set point of the temperature is set by a bottoms analysis controller that maintains constant the methane to ethylene ratio. This tends to produce an ethylene product with constant methane impurity concentration.

The deethanizer is a typical multicomponent column. The top product is a mix of ethylene and ethane, which goes to the reactors, where acetylene is hydrogenated to ethylene. The bottoms product is propylene and heavier components and goes to the depropanizer.

The top control of the deethanizer consists of distillate product flow, which is set in ratio to feed flow. The distillate to feed flow ratio is set by a tray temperature controller 10 to 15 trays from the top. Propylene concentration on top is used by an analysis controller that adjusts the set point of the tray temperature controller to maintain top ethylene concentration at a desired value.

A bottom composition controller, which uses the ethane to propylene ratio to control the ethane impurity in the propylene product, is used to set the boilup to bottoms flow ratio. This ratio is multiplied by the bottoms product flow and the result is the set point of the steam flow controller, which manipulates steam to obtain the desired boilup flow.

The depropanizer is similar to the deethanizer. The top product mix of propylene and propane goes through the methyl acetylene and propadiene reactors, where methyl acetylene and propadiene are hydrogenated The bottoms product is butane and heavier components and is separated in the debutanizer. The control of the depropanizer top is similar to that of the deethanizer, with the exception that the analysis controller uses the ratio of butanes to propane to set the tray temperature controller. The bottoms control consists of a temperature controller two or three trays above the bottom of the column, which sets heat flow.

The debutanizer top control uses pressure to set reflux plus distillate flow. Distillate flow is set by the ratio of distillate to reflux plus distillate, and the ratio is adjusted by the top vapor temperature controller to maintain temperature at a desired value. Reboiler outlet temperature controls the flow of steam to maintain temperature at a desired value.

The cause and effect diagram for the other columns can be built by charting the following cause and effect variables.

EFFECT VARIABLES

- For each column, top and bottoms flow and composition
- Pyrolysis gasoline flow and end point
- Other measured, calculated, and analysis variables
- Overall performance of each column and value added

CAUSE VARIABLES

- For each column, pressure and temperatures
- For each column, feed flow and feed composition
- Other measured and analysis variables

21.6 CAUSE AND EFFECT DIAGRAM FOR ETHYLENE AND PROPYLENE COLUMNS

The effluent from the acetylene converters is fed to the ethylene fractionator for separation. The ethylene product has two impurities, methane and ethane, which are controlled in this column. The methane comes from the bottom of the demethanizer or is introduced with the hydrogen in the converters. Ethylene product is withdrawn as a side product about 10 trays from the top of the column.

Reference 14 provides details of the control for this column. The top gas flow, which is mostly methane and hydrogen, is set in ratio to feed flow. The gas to feed ratio is set by the analysis controller, which is used to control the methane impurity concentration in the ethylene product. Ethylene product flow is also set in ratio to feed flow. The ethylene to feed flow ratio is set by the analysis controller to control the ethane impurity concentration in the ethylene product.

The bottoms product is essentially ethane with ethylene impurity. The heat flow to the column is also controlled in ratio to the feed flow. An analysis controller in the bottoms product stream, measuring ethylene concentration, controls the heat to feed ratio to maintain the ethylene concentration at a desired value. The propylene fractionator receives feed from the methyl acetylene and propadiene converters and splits it into propylene and propane.

The control strategy for the propylene fractionator is similar to that for the ethylene fractionator. The top gas flow, mostly propane, is set in ratio to feed flow. The gas to feed ratio is set by the analysis controller, which is used to control the ethane impurity concentration in the propylene product.

Propylene product flow is also set in ratio to feed flow. The propylene to feed flow ratio is set by the analysis controller to control the propane impurity concentration in the propylene product.

The bottoms product is essentially propane with propylene impurity. The heat flow to the column is also controlled in ratio to feed flow. An analysis controller in the bottoms product stream, measuring propylene concentration, controls the heat to feed ratio to maintain the propylene concentration at a desired value.

The cause and effect diagram for the ethylene and propylene fractionators can be built by charting the following cause and effect variables.

EFFECT VARIABLES

- Methane gas flow and composition
- Ethylene product flow and composition
- Ethane recycle flow
- Ethane gas flow and composition
- Propylene product flow and composition
- Propane recycle flow
- Other measured and analysis variables
- Overall performance of ethylene and propylene fractionators and value added

CAUSE VARIABLES

- For each column, pressure and temperatures
- For each column, feed flow and feed composition
- Other measured and analysis variables

21.7 ETHYLENE PLANT CAUSE AND EFFECT HIERARCHY

A cause and effect diagram such as the one shown in Fig. 15-1 is built for each ethylene plant unit discussed in the preceding sections. The types of charts to be used for the cause and effect variables are selected based on the analysis procedures presented in Chapter 17.

Ethylene Plant Cause and Effect Diagram

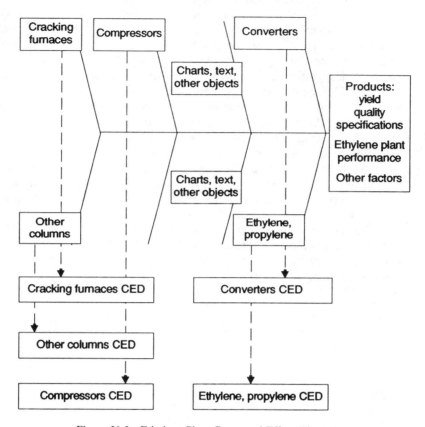

Figure 21-2   Ethylene Plant Cause and Effect Hierarchy.

The unit cause and effect diagrams are organized into the ethylene plant cause and effect hierarchy shown in Fig. 21-2, based on the discussion in Chapter 15. As pointed out in that chapter, when necessary the units can be divided into subunits with their own cause and effect diagrams, or a number of units and subunits can be organized into area cause and effect hierarchies.

The ethylene plant cause and effect hierarchy is then used to monitor and operate the whole ethylene plant with the aim of continuously providing products on specification at minimum cost. It provides all information on cause and effect and aids in the solution of problems.

The same hierarchy can be used by everybody in the ethylene plant, from the plant superintendent to the operators, from the instrumentation and engineering managers to the instrument technicians and the unit engineers, and from information system management to the programmers.

Moving up the hierarchy, information is abstracted to provide what is necessary at that level, no more and no less. Take, for example, the ethylene plant cause and effect diagram of Fig. 21-2, which is the top-level diagram in the hierarchy. For this diagram, the charts and other analysis diagrams and text objects attached to the effect box provide monitoring and all information for all product yields and specifications and for the overall performance of the ethylene plant. The cause boxes provide access to each major unit's cause and effect diagram to zoom in for more detail when necessary. However, the empty cause boxes can be used to attach charts and other objects that furnish abstracted information about the most important cause and effect variables for each unit.

Thus one and all can navigate electronically through all available information from feeds to products in the areas of their responsibility. Moreover, they can also access and navigate through areas that affect them and areas they affect in order to monitor operation and performance continuously and, therefore, strive for continuous improvement and enhanced productivity.

Finally, the ethylene plant cause and effect hierarchy makes everybody continuously sensitive to internal as well as external customers and suppliers. This results in a continuous commitment to maintain and improve customer satisfaction, to improve operation, and thus to improve productivity.

Chapter 22

Approaches to Implementation in the Pulp and Paper Industry

The purpose of this chapter is to use the pulp and paper mill as an example of how to approach the implementation of statistical process control in the pulp and paper industry.

The five major objectives of this chapter are as follows:

1. To provide a process description for the pulp and paper mill.
2. To provide the cause and effect diagram and typical cause and effect variables to be charted for each major unit of the pulp and paper mill.
3. To provide the pulp mill cause and effect hierarchy.
4. To provide the paper mill cause and effect hierarchy.
5. To provide the pulp and paper mill cause and effect hierarchy.

22.1 PULP AND PAPER MILL PROCESS DESCRIPTION

There are many different processes for producing pulp; however, the sulfate pulp mill is discussed here. Reference 18 is the main reference used for both the pulp and paper mill. Figure 22-1 shows a typical sulfate pulp and paper mill diagram.

Logs are collected in the woodyard, where their bark is removed by a me-

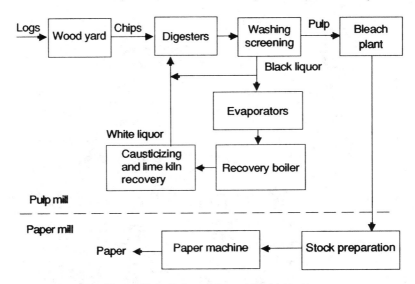

Figure 22-1   Pulp and Paper Mill Diagram.

chanical or hydraulic barker. Debarked logs are fed to the chipper, where they are reduced to small chips less than an inch in size. The chips are stored in separate piles by species, for example, hardwoods (oak) and softwoods (pine).

Wood is mainly composed of cellulose, hemicellulose, and lignin. Cellulose is a white fibrous substance that is insoluble in water and organic solvents and has high tensile strength. It is the principal component of wood, ranging in content from 50% to 55%. Hemicelluloses are carbohydrates in the plant tissues and their content in wood ranges from 15% to 25%. They are also insoluble in water and organic solvents, but behave differently than cellulose toward alkali and acids. Lignin is the natural agent that bonds the wood fibers together. It is the second largest component of wood, ranging from 20% to 30%.

The chips are fed to the sulfate digestion process, which uses sodium hydroxide and sodium sulfide chemicals, referred to as the active alkali. The chips are cooked with these chemicals in order to remove the lignin and separate the wood fibers, while preserving the strength of the cellulose fibers. The brown pulp from the digester is screened and washed and fed to the bleach plant, where it is bleached to the desired brightness and pumped to the paper mill.

In the stock preparation area of the paper mill, the pulp stocks are mixed to achieve the qualities necessary for the specific type of paper to be made. Dyes and other substances are also added to achieve the desired color and physical properties for the paper.

The prepared pulp is spread on the wire of the Fourdrinier machine, which passes over a series of vacuum suction boxes to remove as much water as possible. The wet paper leaves the Fourdrinier machine at 20% consistency.

After leaving the wet end of the paper machine, the wet paper goes to the presses, where it is pressed to remove more water and is then introduced to the steam-heated rolls in the dryer section. The dried paper is then wound in large rolls as it comes off the calenders. The paper may also proceed to further finishing operations depending on final use.

The cooking liquor with the dissolved lignin is separated in the washing stage and is concentrated in the evaporators. Sodium sulfate is added to the concentrated liquor, and the mixture is fed to the recovery boiler, where the organics are burned. The molten smelt from the furnace is dissolved in water and forms the green liquor, which is then causticized using slaked lime in the causticizing plant. The lime mud is recovered and recalcined in the lime recovery area and reused in causticizing. The sodium hydroxide and sodium sulfide solution, which is called white liquor, together with some black liquor recycled from the washers is sent to the digester for reuse in cooking new wood chips.

22.2 CAUSE AND EFFECT DIAGRAM FOR WOODYARD

The operations of the woodyard are mostly mechanical. Therefore, machine monitoring and power use are useful variables to monitor. Another variable to monitor is the power generated by the rejected chips and sawdust burned in the power boiler.

The logs may come from different areas and logging operations. Therefore, the quality of wood from each area needs to be monitored. If the chips are separated into different species, the characteristics of each species should be monitored and used during species changes.

The cause and effect diagram for the woodyard can be built by charting the following cause and effect variables.

EFFECT VARIABLES

- Power used by debarker and chipper
- Power generated by waste chips and sawdust
- Operating condition of mechanical equipment
- Quality characteristics for each wood chip species
- Other measured variables
- Woodyard performance calculation, such as energy used and value added per cord of wood

CAUSE VARIABLES

- Log characteristics
- Log species, size, and type

- Variables monitoring mechanical equipment
- Other measured and analysis variables

The cause and effect variables are charted using the procedures presented in Chapter 17 and are attached to a cause and effect diagram such as the one shown in Fig. 15-1. Real-time SPC monitoring and plant operation then follow as described in Chapter 18.

22.3 CAUSE AND EFFECT DIAGRAM FOR DIGESTERS

Chips can be cooked either continuously (continuous digester) or in batches (batch digester). In either case, the main factors affecting the quality of the pulp produced are the ratio of chemicals to wood, the cooking temperature, and the cooking time. The discussion here is limited to the continuous digester. Similar variables also apply to batch cooking.

Chips from the chip hopper drop into the chip meter, which is followed by a low-pressure feeder that feeds them to the presteaming vessel. From there they fall into the chip chute, which is partially filled with white liquor (cooking chemicals). The chips and liquor are hydraulically conveyed to the top of the digester by the high-pressure feeder.

A production controller sets chipmeter speed to maintain a desired production target. The production of pulp is computed based on chipmeter speed, wood bulk density, pocket fill factor, and yield. The blow flow out of the digester is controlled to maintain the blow flow to wood ratio at a desired value.

White liquor flow is set by the chemical controller to maintain the chemical to wood ratio at a desired value. The ratio is computed using the white liquor flow, the effective alkali test, and the tons of wood per day computed previously.

The black liquor recycle flow is set by the liquor to wood ratio controller to maintain the ratio at a desired value. The ratio is computed using the black liquor flow, the white liquor flow, the steam flows, and the tons of wood per day.

Digester level is primarily controlled by manipulating outlet device speed. It can also set production rate as a secondary variable, while black liquor flow is used for emergency level control.

After a short residence time in the top of the digester, the chemical-impregnated chips enter the top heating zone. In this zone, liquor is circulated from the digester to the heater and back to the digester and raises the temperature as uniformly as possible to a desired value. A temperature controller sets steam flow to control temperature. The chips then enter the bottom heating zone, where the temperature is raised to the desired cooking temperature by again setting steam flow.

At this point, the chips enter the cook zone, where they stay long enough to achieve the desired degree of delignification. The degree of delignification is measured by a titration test in the laboratory and is called the kappa number. This is the pulp quality variable for the discharged pulp.

A model using chemical concentration, temperature at the top of the cook, and cook zone residence time predicts discharged pulp kappa number. The model-predicted kappa number is used by the kappa number controller, which sets the lower heater temperature controller set point to maintain the kappa number at a desired target. The kappa number model is corrected by the laboratory test.

The cause and effect diagram for the digester can be built by charting the following cause and effect variables.

EFFECT VARIABLES

- Blow flow, pulp production, and kappa number
- Other measured and analysis variables
- Overall digester performance and value added

CAUSE VARIABLES

- Chipmeter speed, bulk density, pocket fill factor, and yield
- White liquor flow, effective alkali, and liquor to wood ratio
- Black liquor flow and liquor to wood ratio
- Digester level, pressure, and temperatures
- Model-predicted kappa number
- Other measured, model-predicted, and analysis variables

22.4 CAUSE AND EFFECT DIAGRAM FOR WASHING AND SCREENING

The purpose of screening is to remove knots, uncooked wood particles, and other material from the pulp. This is done using coarse and fine screens before washing.

The consistency of incoming pulp is controlled to a desired value by setting dilution water flow. Washer level is controlled by manipulating the black liquor flow from the brown stock washer to the screens. Vacuum pressure to the screens is controlled by adjusting the pulp flow to the screens.

The pulp is washed in a countercurrent fashion using rotary vacuum washers in series. Incomplete washing makes bleaching difficult. On the other hand, excessive washing increases the cost of chemical recovery.

Each washer level is controlled by manipulating the valve in the filtrate recirculation line. Each filtrate tank level is controlled by throttling the flow from the tank to the washer. Conductivity is used to indicate the concentration of chemicals in the filtrate of the last washer. A conductivity controller sets the hot water flow to the last washer to maintain conductivity at a desired value. The density of the black liquor from the washers that is fed to the evaporators is usually measured using nuclear radiation or refractive index.

The cause and effect diagram for screening and washing can be built by charting the following cause and effect variables.

EFFECT VARIABLES

- Pulp quality to bleach plant
- Filtrate conductivity
- Black liquor density or refractive index
- Other measured and analysis variables
- Overall screening and washing performance

CAUSE VARIABLES

- Pulp flow and consistency
- Screen vacuum pressure
- Washer level
- Hot water flow
- Other measured and analysis variables

22.5 CAUSE AND EFFECT DIAGRAM FOR BLEACH PLANT

Washed pulp is fed to the bleach plant, where it is transformed from the brown pulp from the digester into a bright, white pulp. This is accomplished by a number of bleaching stages that vary from mill to mill. It takes more than a single stage to bleach pulp to the required brightness and at the same time preserve the fiber strength characteristics.

A typical bleach plant is the five-stage chlorination, extraction, chlorine dioxide, extraction, and chlorine dioxide plant, termed the CEDED bleach plant. Each stage generally includes a mechanical mixing tank to mix pulp and chemicals, a reaction tower, and a washer that removes most chemicals and water from the pulp. The pulp flow to the chlorination stage is set by the production controller, which uses flow and consistency to compute the mass flow of pulp.

Clorine gas flow is compensated with pressure and temperature measurements, and the chlorine mass flow is computed. The chlorine flow is then set to maintain the chlorine to pulp mass flow ratio at a desired value. An optical sensor for brightness or an oxidation reduction potential sensor is used just before the mixture of chlorine and pulp enters the tower, and it adjusts the chlorine to pulp mass flow ratio to maintain brightness or oxidation reduction potential at a desired value. The brightness or oxidation reduction potential at the tower outlet is used to set the set point of the inlet brightness or oxidation reduction controller to maintain outlet quality at a desired value.

The extraction stages use sodium hydroxide to extract organic and other matter from the pulp. The flow of sodium hydroxide is adjusted to maintain the mass flow ratio of sodium hydroxide to pulp. The ratio itself is adjusted by a model to maintain the pH at the washer, which is, after the tower, at a desired value. Another model sets the set point of the temperature controller, which in turn sets the steam flow to maintain the inlet temperature at the model-predicted value.

The chlorine dioxide stage uses a model that predicts the required sodium hydroxide flow, tower temperature, and chlorine dioxide flow to maintain at the tower outlet the pH target, chlorine dioxide residual target, and brightness target, respectively. The temperature controller sets steam flow. Laboratory test results are used to calibrate the model.

These are some of the typical control strategies for the bleach plant stages. In addition, many other variables are measured and some of them are also controlled.

The cause and effect diagram for the bleach plant can be built by charting the following cause and effect variables.

EFFECT VARIABLES

- Pulp production rate out of bleach plant, brightness, and strength
- Other measured and analysis variables
- Overall bleach plant performance and value added

CAUSE VARIABLES

- Chlorination-stage pulp rate, chlorine to pulp mass ratio, chlorine flow, chlorine residual, inlet and outlet brightness and/or oxidation reduction potential, temperature, and residence time
- Extraction-stage pulp rate, sodium hydroxide to pulp mass ratio, sodium hydroxide flow, pH, temperature, and residence time
- Chlorine dioxide-stage pulp rate, chlorine dioxide to pulp mass ratio, chlorine dioxide flow, chlorine dioxide residual, pH, sodium hydroxide flow, outlet brightness, temperature, and residence time
- Other measured, model-predicted, and analysis variables

22.6 CAUSE AND EFFECT DIAGRAM FOR EVAPORATOR

The black liquor from the washers at 15% to 20% solids is concentrated to 50% to 55% solids in the multiple-effect evaporator. The multiple effect allows a given quantity of heat to be reused for evaporation many times by boiling less dilute liquid with successively lower boiling points at decreasing vapor pressure with the final effect under vacuum.

A model using black liquor flow and percent of solids sets steam flow. The model is corrected using a percent of solids controller for the thick black liquor product to maintain the percent of solids at a desired set point.

Level controllers maintain the level for each effect, setting the flow out of it. In addition, there are temperature and pressure measurements for each effect. Vacuum pressure at the last stage is controlled using cooling water to the condenser.

The cause and effect diagram for the evaporator can be built by charting the following cause and effect variables.

EFFECT VARIABLES

- Thick black liquor flow and percent of solids
- Overall evaporator performance

CAUSE VARIABLES

- Black liquor feed flow and percent solids
- For each effect level, pressure and temperature
- Other measured and analysis variables

22.7 CAUSE AND EFFECT DIAGRAM FOR RECOVERY BOILER

The main purpose of the recovery boiler is to recover the chemicals from the thick black liquor coming from the evaporator. The black liquor solids is the fuel it burns in order to chemically reduce the recycled chemicals as well as the sodium sulfate makeup, which is mixed with the liquor before it enters the furnace. Steam is a by-product.

The thick black liquor is heated to a desired temperature and sprayed to the hot furnace bed. The flow of liquor is measured, as well as the percent of solids, to obtain the mass flow. Primary airflow is set in ratio to the solids mass flow. The airflow controller in turn manipulates the speed of the forced draft fan. The secondary airflow is also set in ratio to the feed mass flow. Furnace pressure is controlled by setting the speed of the induced draft (exhaust) fan.

Oxygen and other stack analysis measurements are used to adjust the mass to air ratios. Laboratory analysis is sometimes used to analyze the smelt and adjust the air ratios. In addition, the thermal and chemical reduction efficiencies may be computed and used for control monitoring and guidance.

The cause and effect diagram for the recovery boiler can be built by charting the following cause and effect variables.

EFFECT VARIABLES

- Reduction and thermal efficiencies and steam flow
- Smelt analysis variables

- Other measured and analysis variables
- Overall recovery boiler performance and value added

CAUSE VARIABLES

- Black liquor mass flow and liquor mass flow to air ratio
- Pressure and temperatures
- Oxygen and other stack analysis variables
- Other measured, model-predicted, and analysis variables

22.8 CAUSE AND EFFECT DIAGRAM FOR CAUSTICIZING AND LIME RECOVERY

The smelt from the recovery boiler is dissolved and fed to the green liquor storage tank. This liquor is causticized by converting sodium carbonate to sodium hydroxide and lime mud. The white liquor is separated and sent to the digester. The lime mud is pumped to the lime recovery area.

The green liquor flow to the slaker is controlled to a desired set point. Its temperature is also controlled by adjusting the steam flow. The density of the lime mud to the lime kiln is controlled to a desired value by manipulating the flow of water at the suction of the mud transfer pump. Other controls are the level control for various tanks and lime kiln controls.

The cause and effect diagram for causticizing and lime recovery can be built by charting the following cause and effect variables.

EFFECT VARIABLES

- White liquor flow and active alkali
- Causticizing efficiency
- Lime mud density
- Other measured and analysis variables
- Overall causticizing and lime recovery performance

CAUSE VARIABLES

- Green liquor flow and temperature
- Causticizer residence time and mixing
- Lime kiln temperature, fuel to air ratio, and stack analysis
- Other measured and analysis variables

22.9 CAUSE AND EFFECT DIAGRAM FOR STOCK PREPARATION

Pulp from the bleach plant needs to be mixed with other pulp types, dyes, chemical additives, and fillers in appropriate proportions. The stock proportioning system consists of a demand signal that sets the flow rates of each component to be mixed in the mixing chest.

Another preparation step includes the use of the refiner for the mechanical treatment of pulp. The temperature difference across the refiner is maintained at a desired value by moving the refiner plug in or out. The temperature difference controller set point is set by the freeness controller to maintain the refined stock freeness at a desired target.

When dry pulps are used for proportioning, they must be first broken up, dispersed, and diluted. The same is done for recycled paper, referred to as *broke*. The machines used are called repulpers, or pulpers for short. The consistency of repulped pulp is usually controlled to a desired set point by adjusting the dilution flow.

These are some of the typical process units and controls in the stock preparation area. There are others, of course, for specific paper mill requirements.

The cause and effect diagram for stock preparation can be built by charting the following cause and effect variables.

EFFECT VARIABLES

- Paper machine chest and pulp characteristics
- Refined pulp freeness
- Repulped pulp consistency
- Other measured and analysis variables
- Overall stock preparation area performance

CAUSE VARIABLES

- Flows of proportioned components and other characteristics
- Refiner temperature difference and power
- Repulper dilution flow
- Other measured and analysis variables

22.10 CAUSE AND EFFECT DIAGRAM FOR PAPER MACHINE

Prepared stock from the mixing chest is pumped to the machine chest, and from there it goes to the wet end of the Fourdrinier paper machine, where a web is formed from the aqueous suspension of fibers and some water is also removed. Then the paper continues through the dry end, where it is dried, smoothed, and rolled.

Stock from the machine chest is diluted to a desired consistency by a consistency controller that manipulates dilution water flow. From there it passes through a refiner and on to the stuff box. The temperature difference across the refiner sets the refiner plug motor speed, and the temperature difference controller set point is set by couch roll vacuum pressure.

From the stuff box the stock is taken under flow control and pumped to the headbox. Headbox level is controlled by throttling the stock bypass valve. Headbox pressure is controlled by adjusting the vent valve to the atmosphere. The wire speed is also measured and controlled and is used to compute the rush to drag ratio desired between slice jet velocity and wire speed. A bias is then added to it, and the result sets the set point of the headbox pressure controller.

From the head box the pulp is placed onto the wire to form the sheet. The wire is under vacuum in order to remove water as the pulp proceeds to the presses, where more water is removed. Vacuum pressure is controlled by adjusting vent valves. The level of the seal box that collects the water is controlled by manipulating the flow out of it and, if necessary, a water stream into it.

After the press section, the formed paper sheet crosses over to the machine dry end. Multicylinder dryers heated with steam dry the sheet to a desired moisture, and from there it is rolled and sent to further cutting and/or finishing.

The paper sheet moisture and basis weight are measured and controlled by setting steam pressure to the dryers and the stock flow in the wet end of the machine. Other controls consist of maintaining differential pressure across the dryer sections and more sophisticated coordinating controls for paper grade and machine speed changes.

The cause and effect diagram for the paper machine can be built by charting the following cause and effect variables.

EFFECT VARIABLES

- Paper rolls produced, basis weight, moisture, strength, and other quality variables
- Other measured, model-predicted, and analysis variables
- Overall paper machine performance and value added

CAUSE VARIABLES

- Stock flow and consistency
- Headbox level and pressure
- Machine speed and rush to drag ratio
- Pressures, temperatures, and levels
- Other measured, model-predicted, and analysis variables

22.11 PULP AND PAPER MILL CAUSE AND EFFECT HIERARCHY

A cause and effect diagram such as the one shown in Fig. 15-1 is built for each of the pulp and paper mill units discussed in the preceding sections. The types of charts to be used for the cause and effect variables are selected based on the analysis procedures presented in Chapter 17.

The unit cause and effect diagrams are organized into the pulp mill area cause and effect hierarchy shown in Fig. 22-2 and the paper mill area cause and effect hierarchy shown in Fig. 22-3 based on the discussion in Chapter 15. The

Pulp Mill Cause and Effect Diagram

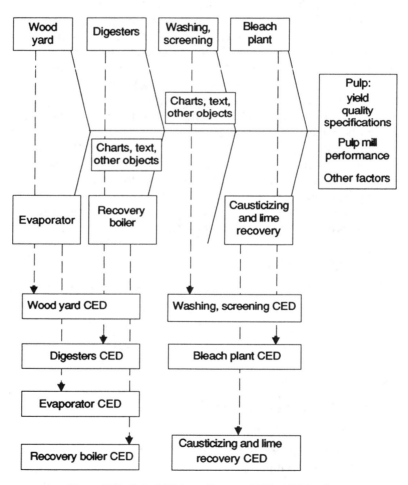

Figure 22-2   Pulp Mill Area Cause and Effect Hierarchy.

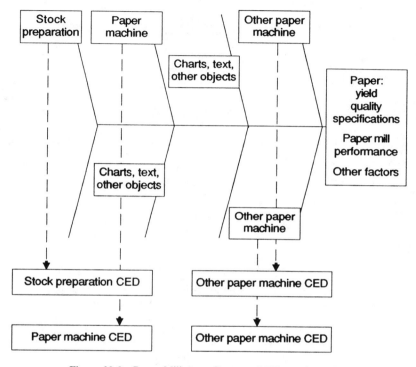

Figure 22-3   Paper Mill Area Cause and Effect Hierarchy.

area hierarchies for the pulp and the paper mill are then used to organize the pulp and paper mill cause and effect hierarchy shown in Fig. 22-4.

The pulp and paper mill cause and effect hierarchy is then used to monitor and operate the whole pulp and paper mill with the aim of continuously providing products on specification at minimum cost. It provides all information on cause and effect and aids in the solution of problems.

The same hierarchy can be used by everybody in the pulp and paper mill, from the plant superintendent to the operators, from the instrumentation and engineering managers to the instrument technicians and the unit engineers, and from information system management to the programmers.

Moving up the hierarchy, information is abstracted to provide what is necessary at that level, no more and no less. Take, for example, the pulp and paper mill cause and effect diagram of Fig. 22-4, which is the top-level diagram in the hierarchy. For this diagram, the charts and other analysis diagrams and text objects attached to the effect box provide monitoring and all information for all product yields and specifications and for the overall performance of the pulp and

Pulp and Paper Mill Cause and Effect Diagram

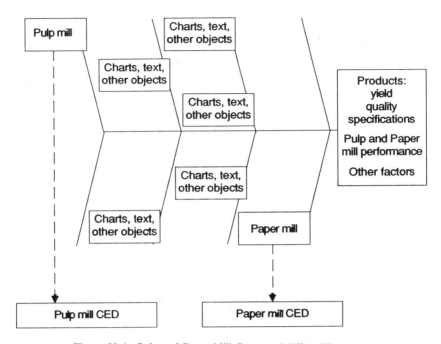

Figure 22-4   Pulp and Paper Mill Cause and Effect Hierarchy.

paper mill. The cause boxes provide access to each pulp mill and the paper mill area cause and effect hierarchy to zoom in for more detail when necessary. The empty cause boxes can be used to attach charts and other objects that furnish abstracted information about the most important cause and effect variables for each area and/or unit.

Thus, one and all can navigate electronically through all available information from feeds to products in the areas of their responsibility. Moreover, they can also access and navigate through areas that affect them and areas they affect in order to monitor operation and performance continuously and, therefore, strive for continuous improvement and enhanced productivity.

Finally, the pulp and paper mill cause and effect hierarchy makes everybody continuously sensitive to internal as well as external customers and suppliers. This results in a continuous commitment to maintain and improve customer satisfaction, to improve operation, and thus to improve productivity.

Chapter 23

Approaches
to Implementation
in the Food Industry

The purpose of this chapter is to use the corn milling process as an example of how to approach the implementation of statistical process control in the food industry.

The three major objectives of this chapter are as follows:

1. To provide a process description for the corn milling process.
2. To provide the cause and effect diagram and typical cause and effect variables to be charted for each major unit of the corn milling process.
3. To provide the corn milling process cause and effect hierarchy.

23.1 CORN MILLING PROCESS DESCRIPTION

Reference 19 is the main reference used for the corn milling process. Figure 23-1 shows a typical corn milling process diagram with the major units and products. Shelled and cleaned corn is conveyed from the storage bins into the steeping tanks, where it is soaked in sulfurous acid for about 40 hours at a temperature of 120° to 130°F. After steeping, the steepwater goes to the steepwater evaporators, where it is concentrated to proper density and sent to feed processing.

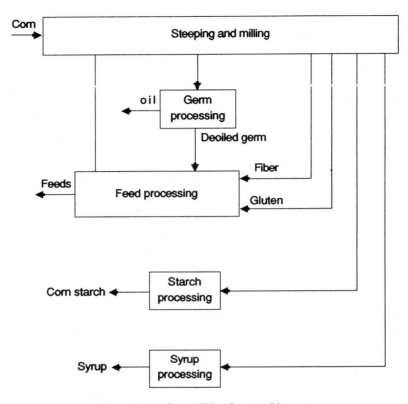

Figure 23-1 Corn Milling Process Diagram.

The softened kernels, on the other hand, go to the degerminating mills, which remove and separate the germs from the kernel. The kernel without the germ is then transported in a water slurry to the grinding mills, where it is ground into a mixture of starch, fiber, and gluten. The mixture is passed through the fiber-removing screens, which separate the fiber.

The slurry of starch and gluten is pumped through the hydrocyclone starch separation units, which separate it into starch and gluten slurries. The gluten is dried and used for animal feed. The starch slurry is used to produce starches and syrups, which are the two major types of products.

23.2 CAUSE AND EFFECT DIAGRAM FOR STEEPING AND MILLING

Steeping, degerminating, grinding, and separation are combined in a single cause and effect diagram, because most of the operation is mechanical and the number of cause and effect variables is not very large. Cleaned corn from storage is brought into the steeping area, where it is placed in large stainless-steel tanks filled with water. Sulfur dioxide is added to the tanks to prevent germination and keep fermentation to a minimum. The control consists of vessel temperature and residence time.

When steeping is completed, the corn goes to the degerminating mills, which tear the softened kernels apart and free the germ. The germ is separated and the remaining mixture is ground in the grinding mills. The ground mixture is separated using hydrocyclones into the overflow gluten slurry and the underflow starch slurry.

The flow and density of the feed to the separators are measured and the flow is controlled. The starch slurry density controller sets starch slurry flow to maintain density at a desired value. Wash water is added to the separators in ratio to the starch flow.

The cause and effect diagram for steeping and milling can be built by charting the following cause and effect variables.

EFFECT VARIABLES

- Germ, glutten, and starch yield and other characteristics
- Other lab analysis and measured variables
- Steeping, milling, and separation performance calculations, such as energy used per unit feed or product and value added per unit feed or product

CAUSE VARIABLES

- Amount of corn feed and characteristics
- Steep vessel temperature and residence time
- Degerminating, grinding, screening, and separation efficiencies
- Milling equipment monitoring variables
- Gluten and starch flows and densities
- Other measured, analysis, and model-predicted variables

The cause and effect variables are charted using the procedures presented in Chapter 17 and are attached to a cause and effect diagram such as the one shown in Fig. 15-1. Real-time SPC monitoring and plant operation then follow as described in Chapter 18.

23.3 CAUSE AND EFFECT DIAGRAM FOR GERM PROCESSING

After dewatering, the germ is dried. The germ dryer uses oil or gas fuel and heats air in a furnace, which is then used to remove moisture from the germ in the dryer. Moist air is then exhausted to the atmosphere.

Typical dryer control consists of an inlet temperature controller that sets fuel flow. The inlet temperature is that of the heated air before it comes in contact with the wet feed at the dryer inlet side. The set point of the inlet temperature controller is set by the outlet temperature controller, which uses exhaust air temperature. In addition, an inferential moisture model is used to maintain constant dryness as a function of load by adjusting the outlet temperature set point. When an on-line germ product moisture sensor is available, it is used to adjust the model-predicted dryness to maintain product moisture at a desired value.

The dried germ is cooled and sold or further processed to extract the oil and refine it. The deoiled germ is then used for animal feed.

Crude oil is mixed with caustic soda and refined in the vessel. Refined oil is then separated by centrifuge as the overflow product. Crude oil flow sets the caustic soda in ratio to it, and a conductivity measurement on dilute solids in the mixing vessel is used for monitoring the operation. The temperature of oil to the centrifuge is controlled by setting the steam flow to the heat exchanger.

Refined oil is then washed with water and separated with a centrifuge into washed oil overflow and soapy water underflow. The temperatures of oil and water are controlled, and water flow is set in ratio to refined oil flow.

Bleach is added to the refined oil and then it is dried in the vacuum dryer, filtered, and deaerated. The dryer temperature is controlled by setting the steam flow into it, while vacuum pressure is controlled by adjusting the water flow to the condenser.

The oil is then hydrogenated by mixing it with hydrogen and catalyst in the converter. It is then filtered, deodorized, and blended into product. Catalyst and hydrogen flow are set in ratio to oil flow. The converter temperature is controlled by adjusting the cooling water flow.

The cause and effect diagram for germ processing can be built by charting the following cause and effect variables.

EFFECT VARIABLES

- Oil yield and characteristics
- Other lab analysis and measured variables
- Germ processing performance calculation, such as energy used per unit feed or product and value added per unit feed or product

CAUSE VARIABLES

- Germ moisture
- Germ dryer temperatures and fuel flow
- Refining vessel oil flow, caustic flow, and conductivity
- Wash water flow and ratio
- Vacuum dryer temperature and pressure
- Converter temperature, oil, catalyst, and hydrogen flows
- Other temperatures

23.4 CAUSE AND EFFECT DIAGRAM FEED PROCESSING

The purpose of feed processing is to take the by-products of the corn milling process and convert them to animal feeds. The main by-products are fiber, gluten, deoiled germ, and steepwater.

The steepwater's 5% to 10% solids is concentrated to 50% solids in multiple-effect evaporators. The multiple effect allows a given quantity of heat to be reused for evaporation many times by boiling less dilute liquid with successively lower boiling points at decreasing vapor pressures, with the final effect under vacuum.

A material and energy balance model using steepwater feed flow and density sets steam flow. The model is corrected using a product density controller to maintain the percent of solids in the product at the desired set point, usually 50%.

Level controllers maintain the level for each effect, setting the flow out of it. In addition, there are temperature and pressure measurements for each effect. Vacuum pressure at the last stage is controlled using cooling water to the condenser.

The fiber and gluten are dewatered and dried to moistures of 8% to 12% water in direct fired and steam tube types of dryers. The control for the direct-fired dryers is similar to that of the germ dryer discussed in the previous section. The steam tube dryer control system uses steam pressure to control steam flow. When available, the moisture sensor measurement is used to set the set point of the pressure controller. In addition, inferential moisture models can also be used.

The cause and effect diagram for feed processing can be built by charting the following cause and effect variables.

EFFECT VARIABLES

- For each evaporator steepwater product flow, percent of solids
- Performance of each evaporator
- Gluten and fiber product, and yield and moisture
- Performance of each dryer
- Overall feed processing performance and value added per unit of product

CAUSE VARIABLES

- For each evaporator feed flow, percent of solids and steam flow
- For each effect level, pressure and temperature
- For each direct-fired dryer fuel flow, temperatures and moisture
- For each steam tube dryer steam flow, pressure and moisture
- Other measured, model-predicted, and analysis variables

23.5 CAUSE AND EFFECT DIAGRAM STARCH PROCESSING

The starch slurry is used to produce modified and unmodified starch and syrups, which are the major products of the corn milling process. The starch is washed in a countercurrent wash system. Control consists of wash water flow set in ratio to slurry flow. The washed starch is then dried to produce the unmodified starch product. The dryer control system is similar to that for the germ dryer when direct-fired dryers are used, but it differs when tunnel dryers are used.

Modified starches are produced by a number of conversion processes. The purpose of conversion is to reduce starch viscosity and to modify properties, such as cooking behavior, cold water solubility, and textural properties.

Conversion is done in batches or continuously. Typical acid conversion controls consist of controlling acid flow in ratio to slurry flow, pH, temperature, and residence time. Dryers are used after conversion and washing.

The cause and effect diagram for starch processing can be built by charting the following cause and effect variables.

EFFECT VARIABLES

- Starch yield, moisture, and other characteristics
- Performance of each conversion process
- Overall starch processing performance and value added per unit of product

CAUSE VARIABLES

- For each conversion process feed flow, acid or enzyme addition, pH, temperature, residence time, and other variables affecting conversion
- For each direct-fired dryer fuel flow, temperatures and moisture
- For each tunnel dryer, zone temperatures and residence time
- Other measured, model-predicted, and analysis variables

23.6 CAUSE AND EFFECT DIAGRAM FOR SYRUP PROCESSING

Corn starch is also the raw material for syrup manufacturing. The starch is converted to syrup using a number of different conversion processes. The acid and the acid–enzyme processes are used to manufacture conventional corn syrups and dextrose. Dextrose is isomerized in reactors that contain enzyme to produce high-fructose corn syrup.

Conversion is done in batches or continuously. Typical acid conversion controls consist of controlling acid flow in ratio to slurry flow, pH, temperature, and residence time.

After conversion the dilute syrup is filtered and centrifuged to remove traces of insoluble protein and starch. It then goes to the evaporator where it is concentrated to the desired density. The control of the syrup evaporator is similar to that for steepwater.

The cause and effect diagram for syrup processing can be built by charting the following cause and effect variables.

EFFECT VARIABLES

- Syrup yield, percent of solids, and other characteristics
- Performance of each conversion process
- Overall syrup processing performance and value added per unit of product

CAUSE VARIABLES

- For each conversion process feed flow, acid or enzyme addition, pH, temperature, residence time, and other variables affecting conversion
- For each evaporator feed flow, percent of solids and steam flow
- For each effect level, pressure and temperature
- Other measured, model-predicted, and analysis variables

23.7 CORN MILLING PROCESS CAUSE AND EFFECT HIERARCHY

A cause and effect diagram such as the one shown in Fig. 15-1 is built for each corn mill process unit discussed in the preceding sections. The types of charts to be used for the cause and effect variables are selected based on the analysis procedures presented in Chapter 17.

The unit cause and effect diagrams are organized into the corn milling process cause and effect hierarchy shown in Fig. 23-2, based on the discussion in Chapter 15. As pointed out in that chapter, when necessary the units can be divided into subunits with their own cause and effect diagrams, or a number of units and subunits can be organized into area cause and effect hierarchies.

The corn milling process cause and effect hierarchy is then used to monitor and operate the whole corn milling process with the aim of continuously providing products on specification at minimum cost. It provides all information on cause and effect and aids in the solution of problems.

The same hierarchy can be used by everybody in the corn mill, from the plant superintendent to the operators, from the instrumentation and engineering managers to the instrument technicians and the unit engineers, and from information system management to the programmers.

Moving up the hierarchy, information is abstracted to provide what is necessary at that level, no more and no less. Take, for example, the corn milling process cause and effect diagram of Fig. 23-2, which is the top-level diagram in the hierarchy. For this diagram, the charts and other analysis diagrams and text objects attached to the effect box provide monitoring and all information for all product yields and specifications and for the overall performance of the corn mill. The cause boxes provide access to each major unit's cause and effect diagram to zoom in for more detail when necessary. However, the empty cause boxes can

Corn Milling Process Cause and Effect Diagram

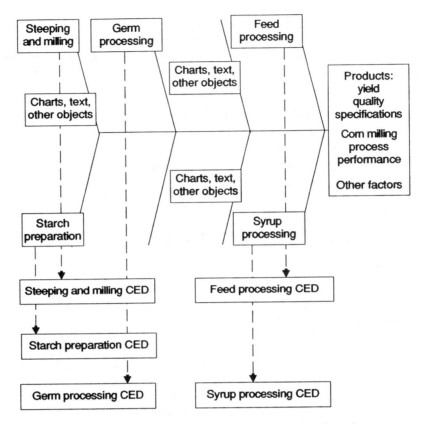

Figure 23-2   Corn Milling Process Cause and Effect Hierarchy.

be used to attach charts and other objects that furnish abstracted information about the most important cause and effect variables for each unit.

Thus one and all can navigate electronically through all available information from feeds to products in the areas of their responsibility. Moreover, they can also access and navigate through areas that affect them and areas they affect in order to monitor operation and performance continuously and, therefore, strive for continuous improvement and enhanced productivity.

Finally, the corn milling process cause and effect hierarchy makes everybody continuously sensitive to internal as well as external customers and suppliers. This results in a continuous commitment to maintain and improve customer satisfaction, to improve operation, and thus to improve productivity.

Chapter 24

Approaches
to Implementation
in the Minerals Industry

The purpose of this chapter is to use the alumina process as an example of how to approach the implementation of statistical process control in the minerals industry.

The three major objectives of this chapter are as follows:

1. To provide a process description for the alumina process.
2. To provide the cause and effect diagram and typical cause and effect variables to be charted for each major unit of the alumina process.
3. To provide the alumina process cause and effect hierarchy.

24.1 ALUMINA PROCESS DESCRIPTION

Reference 20 is the main reference for the alumina process. Figure 24-1 shows a typical alumina process diagram with the major units and products. Bauxite from storage is conveyed to the breaker, which breaks up the ore nodules and removes limestone and other materials. The bauxite then is made into a slurry by mixing it with the recycled caustic liquor. The slurry then enters the digestion unit, which

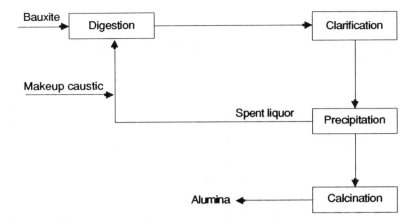

Figure 24-1 Alumina Process Diagram.

consists of three digesters in series, the flash tanks, the blow-off tank, and the liquor heat exchangers.

Digestion of the ore by the caustic solution takes place in the digesters under high temperature and pressure. The digester effluent, composed of sodium aluminate and red mud residue, is cooled as it passes through a series of flash tanks. The flashed steam is then used in the heat exchangers to heat the recycled caustic liquor.

After cooling, the digestion unit effluent enters the clarification unit. First, it goes through the mud settlers where cooked starch is added to aid in flocculation. The underflow of the settlers is washed in a train of washers in a countercurrent fashion to recover caustic. The wash liquid is returned to the last flash tank and the mud is discarded.

The liquor overflow from the settlers goes through pressure filters to remove traces of mud, and then it is cooled by passing through the vacuum flash tanks. From there the liquor goes to the precipitation unit.

The precipitation unit consists of a large number of precipitator tanks and three classifier vessels. Each precipitation tank is filled with liquor and aluminum hydrate seed crystals and is continuously agitated for a number of hours. After the crystals precipitate, each tank is emptied into the primary classifier, which removes the coarse crystals. The overflow of the primary classifier goes to the secondary and tertiary classifiers. The spent liquor overflow of the tertiary classifier is heated and recycled to the digestion unit after the addition of makeup caustic. The fine aluminum hydrate crystals are used for precipitator seed.

The coarse crystals from the primary classifier are washed and filtered to remove water, and from there they go to the calcination unit. In this unit the hydrate crystals are calcined to alumina product in rotary kilns. The product is then cooled and stored for shipping.

24.2 CAUSE AND EFFECT DIAGRAM FOR DIGESTION

The bauxite slurry enters the digesters where it joins the major recycled liquor stream. The alumina reacts with the caustic to produce the soluble sodium aluminate, while the insoluble red mud goes through digestion as suspended solids.

Bauxite to the digester is charged by setting bauxite slurry flow. The slurry flow set point is adjusted to maintain the mass flow of bauxite at a desired value. Digester temperature is controlled by setting the flow of steam injected to the digester. The residence time varies as a function of bauxite and caustic liquor charged to the digester.

Another important variable in extracting the alumina is the ratio of alumina contained in bauxite to caustic in the liquor. This can be computed based on bauxite and liquor analysis and mass flows. The desired alumina to caustic ratio can then be maintained by adjusting bauxite mass flow. If an alumina to caustic measurement is made at the digestion effluent flow, then this measurement can be used to provide feedback adjustment to the bauxite mass flow charged. Because the heat exchangers are fouled over time, their heat transfer efficiency can be monitored by energy balances around each exchanger.

The cause and effect diagram for digestion can be built by charting the following cause and effect variables.

EFFECT VARIABLES

- Digestion effluent, yield, and composition
- Other measured and analysis variables
- Digestion efficiency and other performance calculations, such as energy used per unit feed and value added per unit feed

CAUSE VARIABLES

- Bauxite mass flow and composition
- Caustic flow and composition
- Digester pressure, temperature, steam flow, and residence time
- Heat exchanger heat transfer coefficients, temperatures, and flow
- Other measured, model-predicted, and analysis variables

The cause and effect variables are charted using the procedures presented in Chapter 17 and are attached to a cause and effect diagram, such as the one shown in Fig. 15-1. Real-time SPC monitoring and plant operation then follow as described in Chapter 18.

24.3 CAUSE AND EFFECT DIAGRAM FOR CLARIFICATION

The digestion effluent enters the sand separators in the clarification unit, which remove the coarse material. The sand separator overflow is then fed to the mud settlers, whose main purpose is to remove essentially all the solids in the slurry stream. Starch is added to the settlers to aid flocculation.

The clarity of the liquor overflow is controlled by starch (flocculant) addition. Mud level in the settler is controlled by adjusting the mud underflow. Rake torque is also used to bias the mud underflow adjustment. Mud density and mud alumina and caustic concentration are also used to monitor mud settler and digestion operations.

The concentration of caustic in the washer overflow affects precipitation performance. Therefore, the caustic concentration in the settler feed slurry is controlled by setting dilution flow. This maintains caustic overflow concentration at a desired level.

The mud underflow from the settlers is then washed in countercurrent fashion by a number of washers in series. Washer overflow clarity is monitored by adjusting wash water flow and flocculant addition.

The settler liquor overflow is fed to the presses, where the remaining traces of mud are removed. Flow, differential pressure, and filtrate clarity are used to monitor the operation of each press. This liquor is referred to as pregnant liquor, because it contains the sodium aluminate.

The cause and effect diagram for clarification can be built by charting the following cause and effect variables.

EFFECT VARIABLES

- Press filtrate pregnant liquor clarity, flow, and composition
- Washer mud underflow, density, and composition
- Other measured, model-predicted, and analysis variables
- Clarification efficiency and other performance calculations

CAUSE VARIABLES

- Slurry flow and composition
- Flocculant flow and composition
- Settler overflow clarity and composition
- Settler mud level and rake torque
- Washer overflow, clarity, and composition
- Wash water flow and dilution flows
- Washer temperatures
- Other measured, model-predicted, and analysis variables

24.4 CAUSE AND EFFECT DIAGRAM FOR PRECIPITATION

From the presses the pregnant liquor is cooled by passing through a number of vacuum flash tanks in series. The liquor then goes to the precipitation unit. The precipitation unit has many tanks that are scheduled on a batch basis to convert the sodium aluminate to aluminum hydrate crystals.

Each precipitation tank is filled with pregnant liquor and seed crystals. The important control consists of charging the appropriate amount of seed crystal for the mass of alumina in the pregnant liquor. Precipitation occurs over a period of time while the tank is continuously agitated. The length of the precipitation time of a tank has a direct effect on product yield.

The temperature at which the precipitated slurry is pumped off to the classifiers is another variable that affects precipitation. This temperature is determined by the initial pregnant liquor temperature, the cooling rate, and the precipitation time.

The crystals are then separated in the classifiers, and the coarser crystals are the product fed to the calcination unit, while the finer crystals are recycled for seeding. The classifier overflow spent liquor is recycled to the digestion unit after heating and the addition of makeup caustic. The clarity of the overflow spent liquor is monitored and the charge flow to the classifiers is adjusted to maximize clarity.

The cause and effect diagram for precipitation can be built by charting the following cause and effect variables.

EFFECT VARIABLES

- Coarse and fine hydrate crystals yield
- Spent liquor flow, clarity, and composition
- Other measured and analysis variables
- Precipitation efficiency and other performance calculations

CAUSE VARIABLES

- Pregnant liquor and seed crystal mass charged to each tank
- Precipitation time
- Pump off temperature
- Other measured and analysis variables

24.5 CAUSE AND EFFECT DIAGRAM FOR CALCINATION

The coarse hydrate crystals from precipitation are stored in the hydrate tanks, where water is added. Injection flow to the hydrate tanks is set in ratio to hydrate flow. The ratio is adjusted based on slurry density to the filters.

The slurry is then pumped to hydrate filters that wash the hydrate and remove water. The hydrate is then fed to the rotary kilns that calcine it; that is, they remove the physically and chemically bound water from the hydrate and make the alumina product.

Kiln fuel flow is set in ratio to the mass flow of hydrate and water in the feed. The ratio is adjusted based on product loss on ignition analysis. The kiln airflow is set in ratio to fuel flow, and combustion is monitored using stack gas analysis.

The cause and effect diagram for calcination can be built by charting the following cause and effect variables.

EFFECT VARIABLES

- Alumina product yield and loss on ignition
- Other measured and analysis variables
- Calcination efficiency and other performance calculations

CAUSE VARIABLES

- Slurry density to filters
- Mass of hydrate and water in kiln feed
- Fuel and air flows
- Kiln temperatures
- Stack gas analysis
- Other measured and analysis variables

24.6 ALUMINA PROCESS CAUSE AND EFFECT HIERARCHY

A cause and effect diagram such as the one shown in Fig. 15-1 is built for each alumina process unit discussed in the preceding sections. The types of charts to be used for the cause and effect variables are selected based on the analysis procedures presented in Chapter 17.

The unit cause and effect diagrams are organized into the alumina process cause and effect hierarchy shown in Fig. 24-2, based on the discussion in Chapter 15. As pointed out in that chapter, when necessary the units can be divided into subunits with their own cause and effect diagrams, or a number of units and subunits can be organized into area cause and effect hierarchies.

The alumina process cause and effect hierarchy is then used to monitor and operate the whole alumina plant with the aim of continuously providing products on specification at minimum cost. It provides all information on cause and effect and aids in the solution of problems.

The same hierarchy can be used by everybody in the alumina plant, from the plant superintendent to the operators, from the instrumentation and engineering managers to the instrument technicians and the unit engineers, and from information system management to the programmers.

Moving up the hierarchy, information is abstracted to provide what is necessary at that level, no more and no less. Take, for example, the alumina process cause and effect diagram of Fig. 24-2, which is the top-level diagram in the hierarchy. For this diagram, the charts and other analysis diagrams and text objects

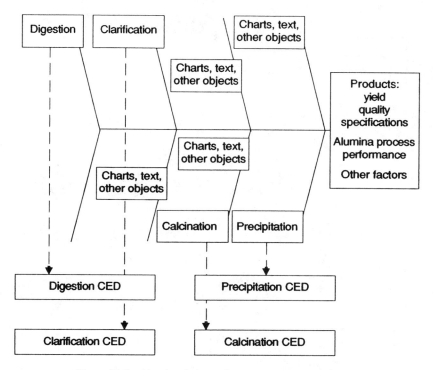

Alumina Process Cause and Effect Diagram

Figure 24-2   Alumina Process Cause and Effect Hierarchy.

attached to the effect box provide monitoring and all information for all product yields and specifications and for the overall performance of the alumina process. The cause boxes provide access to each major unit's cause and effect diagram to zoom in for more detail when necessary. However, the empty cause boxes can be used to attach charts and other objects that furnish abstracted information about the most important cause and effect variables for each unit.

Thus one and all can navigate electronically through all available information from feeds to products in the areas of their responsibility. Moreover, they can also access and navigate through areas that affect them and areas they affect in order to monitor operation and performance continuously and, therefore, strive for continuous improvement and enhanced productivity.

Finally, the alumina process cause and effect hierarchy makes everybody continuously sensitive to internal as well as external customers and suppliers. This results in continuous commitment to maintain and improve customer satisfaction, to improve operation, and thus to improve productivity.

Appendix A

Glossary of Symbols

a, b = coefficients for $Y = aX + b$

$A3$ = skewness

$ANHIGH$ = area of normal above USL

$ANLOW$ = area of normal below LSL

$ANOFSPEC$ = total area of normal off specification

$B4$ = parameter with value given in Table C-3

$C(j)$ = jth subgroup number of defects

$c2$ = parameter with value given in Table C-1

CB = calculated number of defects mean

CI = class interval

CL = calculated central line

CLR = calculated central line range

CLS = calculated central line sigma

$CLXB$ = calculated central line xbar

CP = inherent process capability

CPK = process capability worst case

CPL = inherent process capability lower

CPU = inherent process capability upper

CR = capability ratio

d2 = parameter with value given in Table C-1

D4 = parameter with value given in Table C-2

Gamma = kurtosis

HA = high alarm

h1 = decision interval for cumulative sum high

h2 = decision interval for cumulative sum low

K = process mean versus specification midpoint

k1 = slack value for cumulative sum high

k2 = slack value for cumulative sum low

K1 = parameter for the $K1$-sigma limits (default $K1 = 3.0$)

KC1 = high gain of the CUSUM controller for *SH*, which can have positive and negative values

KC2 = low gain of the CUSUM controller for *SL*, which can have positive and negative values

KPAR = parameter for the Optimum Setpoint Controller

LA = low alarm

LCL = calculated lower control limit

LCLR = calculated lower control limit range

LCLS = calculated lower control limit sigma

LCLXB = calculated lower control limit xbar

LCL(j) = *j*th subgroup lower control limit

LSL = lower specification limit

LSLR = lower specification limit range

LSLS = lower specification limit sigma

LSLXB = lower specification limit xbar

M = skip size

M(j) = current value of the manipulated variable

MINSIG = minimum sigma value

N = subgroup size

ND(j) = *j*th subgroup number defective

NPB = calculated number defective mean

NS = number of subgroups

OCB = official number of defects mean

OCL = official central line

OCLR = official central line range

OCLS = official central line sigma

OCLXB = official central line xbar

OLCL = official lower control limit

OLCLR = official lower control limit range

OLCLS = official lower control limit sigma

OLCLXB = offical lower control limit xbar

ONPB = official number defective mean

OPB = official fraction defective mean

ORB = official range mean

OSB = official sigma mean

OSIGP = official sigma prime

OSIGR = official range sigma

OSIGS = official sigma sigma

OSIGXB = official xbar sigma

OSPT = optimum set point

OUB = official defects per unit mean

OUCL = official upper control limit

OUCLR = official upper control limit range

OUCLS = official upper control limit sigma

OUCLXB = official upper control limit xbar

OXB = official mean

OXBB = official xbar mean

PB = calculated fraction defective mean

pi = 3.14159

$P(j)$ = jth subgroup fraction defective

RB = calculated range mean

rho = correlation coefficient

$R(j)$ = jth subgroup range

S = sample sigma, calculated using $NS - 1$ samples

SB = calculated sigma mean

$SH(j)$ = cumulative sum high, above target, up to and including subgroup j

sigma = standard deviation

SIGP = sigma prime, calculated using NS samples

SIGR = calculated range sigma

SIGS = calculated sigma sigma

SIGXB = calculated xbar sigma

size = sample size

$S(j)$ = jth subgroup sigma

SL = specification limit

SL(j) = cumulative sum low, below target, up to and including subgroup *j*

SZ(j) = *j*th subgroup size

TAR = target

TARR = target range

TARS = target sigma

TARXB = target xbar

TD = time delay

T0 = period of oscillation

TSS = time to reach steady state

UB = calculated defects per unit mean

UCL = calculated upper control limit

UCLR = calculated upper control limit range

UCLS = calculated upper control limit sigma

UCLXB = calculated upper control limit xbar

U(j) = *j*th subgroup defects per unit

USL = upper specification limit

USLR = upper specification limit range

USLS = upper specification limit sigma

USLXB = upper specification limit xbar

W(i) = weighting coefficient *i*

XB = calculated mean

xbar = mean

XBB = calculated xbar mean

XB(j) = *j*th subgroup mean

X(i) = *i*th sample

X(i, j) = *i*th value of the variable in subgroup *j*

XMAX(j) = *j*th subgroup maximum value

XMIN(j) = *j*th subgroup minimum value

Y(j) = *j*th subgroup normalized deviation from target

Appendix B

Glossary of Terms

Assignable causes (special causes) Assignable causes are those due to which a variable is not in statistical-control. They consist of changes and variations of systematic variables, such as feed flow and composition, temperatures, pressures, and residence times for continuous variables, and faulty setup, defective raw material, and operator error in manufacturing processes.

Autocorrelation coefficient A quantitative measure of the correlation of a variable against itself for a given time delay. The autocorrelation coefficient ranges in value from $+1.0$ to 0. When the time delay is 0, the autocorrelation coefficient is 1.0, meaning that the variable is correlated to itself 100% (maximum correlation). As the time delay increases, the autocorrelation decreases toward 0, but it usually settles to a minimum value above 0.

Average See *mean*.

Cause and effect diagram The cause-and-effect (fishbone) diagram was developed by Kaoru Ishikawa in his work with quality circles in Japan. It is used to document and classify the relationships between quality effects and their causes. The head of the fish is a box indicating a specific effect, while the entries on the bones indicate the causes. The CE diagram not only documents the effect and its causes but it is also used as an active information access mechanism. This is achieved by linking charts and text files to each cause or effect box, which is an active screen area selected by touch or a pointing device. Also, boxes linked to monitored charts show the out-of-statistical-control alarm status of the charts via background color.

C chart A plot of number of defects together with the mean and the upper and lower control limits. It is used with constant unit size, for which the control limits are also constant.

Central line The line drawn at the center of a control chart that represents the mean value of the charted variable.

Central tendency The value about which a group of values is clustered. It can be the mean, median, mode, or other measure of their distribution.

Common cause A variable is in statistical control when all variation due to assignable causes has been eliminated. The remaining variation is totally random and it is the result of common causes.

Control limits The limits represent the values above and below the mean within which the values of a variable remain most of the time, for example, the ± 3-sigma limits from the mean.

CP Inherent capability of process as manifested by a measured variable, which is the specification spread divided by the 6-sigma variation. When $CP > 1.0$, the process is capable.

CPK Process capability based on worst-case view of the data. When $CPK > 1.0$, the process is capable.

CR Capability ratio, which is the inverse of CP. When $CR < 1.0$, the process is capable.

Cross-correlation coefficient A quantitative measure of the correlation between two variables. The cross-correlation coefficient ranges in value from -1.0 to $+1.0$. Zero means that the two variables do not affect one another. A value of $+1.0$ means that there is a strong positive correlation between them. That is, when one increases, the other increases also. A value of -1.0, on the other hand, means that there is a strong negative correlation between them. That is, when one increases, the other decreases. Values between 0 to 1.0 and 0 to -1.0 indicate that there may be positive or negative correlation, with confidence increasing as the values approach 1.0 or -1.0.

Cumulative sum chart A plot of the cumulative deviation of the subgroup mean from the target value. The control limits are in the form of a V-mask that provides a two-sided decision criteria similar to the 3-sigma limits of the xbar chart.

CUSUM chart A plot of the cumulative deviation of the subgroup mean from the target value divided by the subgroup sigma. The control limits are the two decision interval parameters. It can be used instead of the cumulative sum chart, because it is iterative and thus easier for real-time monitoring; also, because the deviation is normalized with respect to sigma it is easier to setup and understand.

Histogram A frequency distribution of a group of values for a given variable, with their mean and sigma used to plot the superimposed normal distribution curve.

Individual Individual or single values of a variable as opposed to subgroups of values.

Individuals chart A plot of a group of individual measurements together with their mean and the upper and lower control limits.

In statistical control A variable is in statistical control or statistically stable when all variation due to assignable causes has been eliminated.

K Process mean versus specification mean. If $K > 0$, the process mean is above the specification mean. If $K < 0$, the process mean is below the specification mean. If $K = 0$, the process mean is equal to the specification mean.

Kurtosis Kurtosis is the degree of peakedness of a frequency distribution, usually in relation to the normal distribution. A kurtosis < 0 indicates a thin distribution with a relatively high peak, which is called leptokurtic. Kurtosis > 0 indicates a distribution that is wide and flat-topped, which is called platykurtic. The normal distribution that falls between the two is called mesokurtic, with kurtosis $= 0$.

Mean The mean is the sum of all values in a group divided by their number.

NP chart A plot of number (of items) defective together with the mean and the upper and lower control limits. It is used with constant subgroup size, for which the control limits are also constant.

Official values Official values for the mean and/or sigma for a variable are computed from a data set that has been chosen as a representative operation to judge the future.

Pareto diagram A plot of causes of rejection for a product in order of priority. The number of occurrences as well as the percent of contribution for each cause are plotted. Moreover, the sum of occurrences over a given period can be plotted instead.

P chart A plot of fraction (of items) defective together with the mean and the upper and lower control limits. The control limits vary as the subgroup size varies.

Process The combination of people, equipment, raw materials, environment, and methods that effects chemical and/or physical changes or other appropriate changes to provide useful products and/or services.

Range The difference between the maximum and minimum values in a given subgroup.

Real-time database Refers collectively to all the distributed databases that collect and store the value, date/time, and status for all plant variables from the current (present) time to some time in the past. It is important to emphasize that the real-time database is always tied to the current time. Thus the SPC charts and other analysis tools, when requested, automatically access data from the current time backward. Thus timely action can be taken to improve operation.

Sample time The time between successive samples of a given variable.

Scatter diagram A plot of one variable against another or itself, with optional time delay, thus providing the ability to compute the cross-correlation between two variables or a variable's autocorrelation for the given time delay.

Sigma The standard deviation.

Skewness Skewness is the degree of asymmetry of a frequency (density) distribution. When the frequency curve of a distribution has a longer tail to the right of the central maximum as compared to the left, the frequency distribution is said to be skewed to the right or to have positive skewness. When the reverse is true, the frequency distribution is said to be skewed to the left or to have negative skewness. A skewness of 0 indicates no skewness.

SPC Statistical process control is the application of statistical tools during the process of making products and/or providing services. SQC is the more general term, although many times they are used synonymously.

Specifications The required properties for product acceptance, for example, that the 6-sigma spread be within the specification spread and the mean be equal to the required target.

SQC Statistical quality control is the application of statistical principles in order to maintain an acceptable product quality on a continuous basis.

Standard deviation A measure of dispersion or variation of a group of values for a given variable about their mean.

Subgroup A number of values for a given variable obtained at about the same time or successively within a reasonable time span. Subgroup members should be as much alike as possible, while the subgroups should be as different from one another as possible.

***U* chart** A plot of defects per unit together with the mean and the upper and lower control limits. The control limits vary as the unit size per subgroup varies.

Xbar See *mean.*

Xbar and range chart A double chart. The upper is a plot of subgroup means together with their mean value and the upper and lower control limits. The lower is a plot of subgroup ranges together with their mean and the upper and lower control limits.

Xbar and sigma chart A double chart. The upper is a plot of subgroup means together with their mean value and the upper and lower control limits. The lower is a plot of subgroup sigmas together with their mean and the upper and lower control limits.

Yield The products of a given process.

Appendix C

Tables

TABLE C-1 FACTORS FOR ESTIMATING POPULATION SIGMA (*SIGP*) FROM THE RANGE MEAN (*RB*) AND THE SIGMA MEAN (*SB*)*

Subgroup Size N	$d2 = \dfrac{RB}{SIGP}$	$c2 = \dfrac{SB}{SIGP}$
2	1.128	0.5642
3	1.693	0.7236
4	2.059	0.7979
5	2.326	0.8407
6	2.534	0.8686
7	2.704	0.8882
8	2.847	0.9027
9	2.970	0.9139
10	3.078	0.9227
11	3.173	0.9300
12	3.258	0.9359
13	3.336	0.9410
14	3.407	0.9453
15	3.472	0.9490
16	3.532	0.9523
17	3.588	0.9551
18	3.640	0.9576
19	3.689	0.9599
20	3.735	0.9619
21	3.778	0.9638
22	3.819	0.9655
23	3.858	0.9670
24	3.895	0.9684
25	3.931	0.9696
30	4.086	0.9748
35	4.213	0.9784
40	4.322	0.9811
45	4.415	0.9832
50	4.498	0.9849
55	4.572	0.9863
60	4.639	0.9874
65	4.699	0.9884
70	4.755	0.9892
75	4.806	0.9900
80	4.854	0.9906
85	4.898	0.9912
90	4.939	0.9916
95	4.978	0.9921
100	5.015	0.9925

* Adapted from E. L. Grant and R. S. Leavenworth, *Statistical Quality Control*, 6th ed. New York: McGraw-Hill Book Company, 1988. Reproduced with permission of McGraw-Hill, Inc.

TABLE C-2 FACTOR FOR DETERMINING THE CONTROL LIMITS OF THE RANGE CHART FROM THE RANGE MEAN *RB**

Subgroup Size N	$D4$
2	3.27
3	2.57
4	2.28
5	2.11
6	2.00
7	1.92
8	1.86
9	1.82
10	1.78
11	1.74
12	1.72
13	1.69
14	1.67
15	1.65
16	1.64
17	1.62
18	1.61
19	1.60
20	1.59

* Adapted from E. L. Grant and R. S. Leavenworth, *Statistical Quality Control,* 6th ed. New York: McGraw-Hill Book Company, 1988. Reproduced with permission of McGraw-Hill, Inc.

TABLE C-3 FACTOR FOR
DETERMINING THE CONTROL
LIMITS OF THE SIGMA CHART
FROM THE SIGMA MEAN
(SB)*

Subgroup Size N	$B4$
2	3.27
3	2.57
4	2.27
5	2.09
6	1.97
7	1.88
8	1.81
9	1.76
10	1.72
11	1.68
12	1.65
13	1.62
14	1.59
15	1.57
16	1.55
17	1.53
18	1.52
19	1.50
20	1.49
21	1.48
22	1.47
23	1.46
24	1.45
25	1.44
30	1.40
35	1.37
40	1.34
45	1.32
50	1.30
55	1.29
60	1.28
65	1.27
70	1.26
75	1.25
80	1.24
85	1.23
90	1.23
95	1.22
100	1.21

* Adapted from E. L. Grant and R. S. Leavenworth, *Statistical Quality Control*, 6th ed. New York: McGraw-Hill Book Company, 1988. Reproduced with permission of McGraw-Hill, Inc.

Appendix D

References

1. W. Edwards Deming, *Out of the Crisis*. Cambridge, MA: MIT Center for Advance Engineering Study, 1986.
2. Paul C. Badavas and Albert D. Epperly, "Statistical Process Control Embedded in Open Industrial Systems," *ISA Transactions*, vol. 30, no. 1 (January 1991). Also presented at *ISA/88 International Conference and Exhibit*, October 16–21, 1988, Houston, TX.
3. Albert D. Epperly and Paul C. Badavas, "Statistical Process Control in an Open Industrial System," *1988 National Petroleum Refiners Association Computer Conference*, October 30–November 2, 1988, Pittsburgh, PA.
4. Eugene L. Grant and Richard S. Leavenworth, *Statistical Quality Control*, 6th ed. New York: McGraw-Hill Book Company, 1988.
5. Western Electric Company, Inc., *Statistical Quality Control Handbook*, 2nd ed. Charlotte, NC: Delmar Printing Company, 1956.
6. George E. P. Box, W. G. Hunter, and J. S. Hunter, *Statistics for Experimenters*. New York: John Wiley & Sons, Inc., 1978.
7. Dana L. Ulery, "Software Requirements for Statistical Quality Control," *ISA, 1986, International Conference and Exhibit*, Houston, TX, October 13–16, 1986, 821–827.
8. James M. Lucas and Ronald B. Crosier, "Fast Initial Response for CUSUM Control Schemes: Give Your CUSUM a Head Start," *Technometrics*. vol. 24, no. 3 (August 1982), 199–205.

9. Kaoru Ishikawa, *Guide to Quality Control*. Tokyo, Japan: Asian Productivity Organization, 1976.

10. Albert D. Epperly and Paul C. Badavas, "Statistical Quality Control, Statistical Process Control: Part 1, Quality Improvement; Part 2, Participative Management; Part 3, Statistical Tools; Part 4, Problem Solving," *The Foxboro Company State-of-the-Art Newsletter*, sec. 3, nos. 5–8, October–December 1987 and January 1988.

11. Ron Gibson, "Quality and the Changing Role of Management," *Machine and Tool BLUE BOOK* (December 1986), Q-10 to Q-12.

12. Carroll J. Ryskamp and Paul C. Badavas, "Profitable and Stable Control for Alcohol Separation in the Presence of Water," *Proceedings of the 1987 American Control Conference*, Minneapolis, MN, June 10–12, 1987.

13. James H. Gary and Glenn E. Handwerk, *Petroleum Refining Technology and Economics*, 2nd ed. New York: Marcel Dekker, Inc., 1984.

14. F. G. Shinskey, *Distillation Control*, 2nd ed. New York: McGraw-Hill Book Company, 1984.

15. Carroll J. Ryskamp, "New Strategy Improves Dual Composition Column Control," *Hydrocarbon Processing*, June 1986.

16. Carroll J. Ryskamp, Noal F. McGee, and Paul C. Badavas, "Better Alkylation Control," *Hydrocarbon Processing*, November 1986.

17. 1985 Petrochemical Handbook, *Hydrocarbon Processing*, November 1985.

18. John R. Lavigne, *Instrumentation Applications for the Pulp and Paper Industry*. San Francisco: Miller Freeman Publications, 1979.

19. "Products of the Corn Refining Industry in Food," *Seminar Proceedings*, Corn Refiners Association, Inc., Washington, DC, May 9, 1978.

20. T. H. Arnold and N. P. Chopey, "New Ideas Refresh Alumina Process," *Chemical Engineering*, November 1960.

Index